Pollution Control in East Asia

Lessons from the Newly Industrializing Economies

Michael T. Rock

RESOURCES FOR THE FUTURE — WASHINGTON, DC, USA
INSTITUTE OF SOUTHEAST ASIAN STUDIES — SINGAPORE

Printed in the United States of America

An RFF Press book
Published by Resources for the Future
1616 P Street, NW, Washington, DC 20036–1400

A copublication of Resources for the Future (www.rff.org) and the Institute of Southeast Asian Studies (www.iseas.edu.sg/pub.html).

Distributed in the ASEAN countries, Japan, Korea, Taiwan, Hong Kong, India, and Australia/New Zealand by the Institute of Southeast Asian Studies, 30 Heng Mui Keng Terrace, Singapore 119614 (ISBN 981-230-163-1).

Library of Congress Cataloging-in-Publication Data
Rock, Michael T.
Pollution control in East Asia : lessons from newly industrializing economies / Michael T. Rock
p. cm.
Includes bibliographical references and index.
ISBN 1-891853-47-3 (lib. bdg.) — ISBN 1-891853-48-1 (pbk.)
1. Factory and trade waste—East Asia—Management—Case studies. 2. Environmental policy—East Asia—Case studies. 3. Factory and trade waste—Asia, Southeastern—Management—Case studies. 4. Environmental policy—Asia, Southeastern—Case studies. 5. Factory and trade waste—Developing countries—Management—Case studies. 6. Environmental policy—Developing countries—Case studies. I. Title
TD897.8.E18 R63 2001
363.73′5′0959—dc21 2001051090

f e d c b a

The paper in this book meets the guidelines for permanence and durability of the Committee on Production Guidelines for Book Longevity of the Council on Library Resources.

The text of this book was designed and typeset by Betsy Kulamer in Trump Medieval and ITC Franklin Gothic. It was copyedited by Alfred F. Imhoff. The cover was designed by Rosenbohm Graphic Design.

ISBN 1–891853–47–3 (cloth) ISBN 1–891853–48–1 (paper)

About Resources for the Future and RFF Press

Resources for the Future (RFF) improves environmental and natural resource policymaking worldwide through independent social science research of the highest caliber.

Founded in 1952, RFF pioneered the application of economics as a tool to develop more effective policy about the use and conservation of natural resources. Its scholars continue to employ social science methods to analyze critical issues concerning pollution control, energy policy, land and water use, hazardous waste, climate change, biodiversity, and the environmental challenges of developing countries.

RFF Press supports the mission of RFF by publishing book-length works that present a broad range of approaches to the study of natural resources and the environment. Its authors and editors include RFF staff, researchers from the larger academic and policy communities, and journalists. Audiences for RFF publications include all of the participants in the policymaking process—scholars, the media, advocacy groups, nongovernmental organizations, professionals in business and government, and the general public.

About the Institute of Southeast Asian Studies

The **Institute of Southeast Asian Studies** (ISEAS) was established in Singapore as an autonomous organization in 1968. It is a regional research center for scholars and other specialists concerned with modern Southeast Asia, particularly the multifaceted problems of stability and security, economic development, and political and social change.

Resources for the Future

Contents

For Maggie,
Oh how far we have together come in such little time,
yet we have miles to go before we sleep.

Preface

Work on this project began somewhat accidentally nearly six years ago, when I had a firsthand opportunity to examine the evolution of industrial pollution management in Taiwan, China. (Through the remainder of this book, the economy of "Taiwan, China" will be referred to as "Taiwan.") At the time, there was very little published literature on this topic, and what did exist was quite critical of the government's environmental policies. Because I was familiar with the political-economy literature on Taiwan's economic development, I wondered whether the government's institutions of industrial policy were subverting the development of more effective pollution management policies or whether those institutions were becoming actively engaged in pollution management. This led me to interview individuals in Taiwan's Environmental Protection Administration and in its institutions of industrial policy, particularly individuals in the Council of Economic Planning and Development, Industrial Development Bureau of the Ministry of Economic Affairs, and Industrial Technology Research Institute.

As a result of my examination of data on ambient air and water quality, I came to the conclusion that Taiwan's Environmental Protection Administration was beginning to successfully clean up industrial pollution and that the institutions of industrial policy were becoming actively engaged in industrial pollution control. I shared an early draft of the paper I wrote at the time with two colleagues, Robert Wade and Jomo K.S.; both offered substantial encouragement. Without their encouragement, this book might never have come to fruition. I am deeply indebted to both.

An updated and revised version of my Taiwan essay appears as Chapter 3 in this volume. This essay was initially published as "Toward More Sustainable Development: The Environment and Industrial Policy in Taiwan," in *Development Policy Review* (Rock 1996b) and subsequently republished in full under the same title in Angel and Rock (2000). Permission to use this material has been granted by Basil Blackwell Publishers and Greenleaf Publishing Limited. After the publication of this Taiwan essay, I completed, with several other colleagues, empirical studies of the determinants of plant-level expenditures on pollution abatement in Indonesia and South Korea. Both of those studies demonstrated that manufacturing plants in each economy were investing in pollution control because of pressure from regulators, communities, and buyers of their products. These studies were consistent with a growing body of empirical work on the determinants of pollution abatement in a number of developing economies in Asia (available at a World Bank website, http://www.worldbank.org/nipr).

Taken together, this body of work and my Taiwan essay convinced me that I should try to extend my Taiwan study to several other East Asian newly industrializing economies. The outcome of that effort is this book on the political economy of industrial pollution management in six East Asian newly industrializing economies—China, Indonesia, Malaysia, Singapore, Taiwan, and Thailand.

My research methodology in the book and in each chapter is quite simple. To begin with, I rooted the analysis in the political economy of policymaking in each case study. The rich literature on this topic was enormously helpful in completing this task. I supplemented this with the rather thin literature on the evolution of each economy's command-and-control environmental regulatory agency. I then turned to the all too few empirical studies on industrial pollution management in each economy. All of this was further supplemented by data gathering on ambient air and water quality in each economy and by in-depth interviews with knowledgeable government officials, researchers, and those in nongovernmental environmental organizations in each economy. I am indebted to the numerous individuals in each economy who agreed to talk with me; most asked that I respect their confidentiality, which I did by referring to interviews in various places in the text without naming names. Their candor was essential in developing this book. The above literature and data, along with the results from my interviews, were then woven together to tell a story about the political economy of industrial pollution management in each case.

Drafts of each case study were shared with a small group of experts from each economy. I am indebted to Jomo K.S. and Chang Yii Tan for comments on Malaysia. And I am indebted to Ooi-Giok Ling for her very helpful comments on the chapter on Singapore and to Watcharee Limanon for her comments on Thailand. Frequent conversations with Shakeb Afsah helped sharpen and improve the chapter on Indonesia. I am also indebted to Li-Fang Yang and Robert Wade for their comments on the chapter on Taiwan. The chapter on China's public disclosure program on the environmental performance of China's cities would have been impossible without the help of Chonghua Zhang, director, China Field Office, Winrock International, and Fei Yu, vice director, Division of Urban and Industrial Management, State Environmental Protection Administration, People's Republic of China. I am deeply indebted to both for all the time they spent with me as we traveled in China and for the help and insights they provided as we developed the chapter on China. I am also indebted to Andrew MacIntyre, Robert Wade, and several anonymous reviewers for their very helpful comments on a draft of the book.

Finally, I would like to thank my friends at Resources for the Future (RFF). I am especially indebted to Michael Toman, Senior Fellow at RFF, for encouraging me to submit early draft chapters to RFF for consideration. I would also like to thank Don Reisman, Director of the RFF Press, for all his support and help. Working with him and his staff, particularly Rebecca Henderson, was both easy and enjoyable.

Michael T. Rock
Alexandria, Virginia

1

Pollution Management Strategies in East Asian NIEs

The literature on the economic and political development of East Asia's newly industrializing economies (NIEs) is vast, rich, and highly contentious. Unfortunately, the same cannot be said for the literature on industrial pollution management in these economies. (These NIEs include China, Hong Kong, Indonesia, Malaysia, South Korea, Taiwan[1], Thailand, and Singapore.) To begin with, there are relatively few studies of industrial pollution management in the East Asian NIEs, and those that do exist are largely descriptive or highly critical. Several trace the evolution of environmental laws and regulations (O'Connor 1994) and analyze the sources of weak industrial pollution management systems (World Bank 1993, 1994a, 1994b). Others examine the reasons for and environmental consequences of these governments' apparent "grow first, clean up later" environmental strategies (Chun-Chieh 1994; Smil 1993; Eder 1996; Bello and Rosenfeld 1992; Komin 1993). Still others describe popular responses to growing urban environmental stress (Cribb 1990; Lee and So 1999).

The picture presented by this body of work is one of urban environments that are rapidly deteriorating in the face of high-speed industrial growth. It is important to note that this picture is not entirely inaccurate. Some of the regions' megacities—Bangkok, Beijing, Jakarta, and Shanghai—are among the dirtiest in the world. At least one analyst has contended (Lohani 1998), and Table 1-1 agrees, that Asia is the dirtiest place in the world. Moreover, a legacy of apparent environmental neglect in some places imposes large envi-

Table 1-1. Environmental Conditions in Asia, 1991–1995

Condition	*Asia*	*Africa*	*Latin America*	*OECD countries*[a]	*World*
Air pollution (μg/m³)					
Particulates	248	29	40	49	126
Sulfur dioxide	0.023	0.015	0.014	0.068	0.059
Water pollution (μg/l)					
Suspended solids	638	224	97	20	151
Level of biological oxygen demand	4.8	4.3	1.6	3.2	3.5

Note: All numbers are averages. Units of measure: "μg/m³" is micrograms per cubic meter; "μg/l" is micrograms per liter.

[a]The countries that belong to the Organisation for Economic Co-operation and Development.

Source: Lohani 1998.

ronmental health costs on urban residents (World Bank 1992a, 1997a, 1997b).

Nonetheless, this book will argue that this extremely negative picture is seriously incomplete. The governments of several East Asian NIEs, notably Malaysia and Singapore, have been effectively cleaning up the environment during urban industrial growth—or "cleaning up while growing." Several other governments, such as those of South Korea and Taiwan, have followed "grow first, clean up later" environmental strategies but are now amassing admirable records of industrial pollution control. Still other governments, such as those of China and Indonesia, have made important (if partial and incomplete) environmental improvements. In fact, of the six East Asian NIEs whose case studies form the basis of this book, only Thailand has failed to achieve at least one significant success in industrial pollution management. But these accomplishments continue to remain more or less hidden by a macroeconomic view (e.g., Table 1-1) that portrays unabated environmental degradation.

Things are not much better at the microeconomic level. Of the all too few studies that attempt to rigorously assess the effectiveness of industrial pollution management policies in the East Asian NIEs, findings remain, at best, mixed. Neither Panayotou (1998) nor Spofford and others (1996) found much evidence that environmental policies in China had much effect on the behavior of polluters or on ambient environmental quality, yet Wang and Wheeler (1996, 1999) concluded that wastewater emissions charges in China have had a significant effect on industrial water emissions. (These emissions charges are levies on wastewater emissions that exceed emissions standards.) Similarly, even though the World Bank (1994a) has generally concluded that Indonesia's environmental impact management agency, BAPEDAL, has had little effect, it credits two of its innovative programs—a Clean Rivers program (World Bank 1994a, 133) and a public disclosure program, PROPER (Afsah and Vincent 2000)—with making a difference.

Fortunately, a relatively smaller but growing body of empirical work now suggests that East Asian NIEs' pollution management programs are having at least modest effects. Wang and Wheeler (1996, 1999), Afsah and Vincent (2000), and Hettige and others (1996) have demonstrated that manufacturing plants in several NIEs are making significant investment in pollution control.[2] Aden and Rock (1999) have demonstrated how a nascent city-level environmental management agency in Indonesia used a traditional, but limited, monitoring and enforcement program to get manufacturing plants to begin to invest in pollution abatement. Aden and others (1999) have also shown how a much tougher traditional monitoring and enforcement program in South Korea led manufacturing plants to significantly invest in pollution abatement. And Vincent and others (2000) have demonstrated how the Department of the Environment in Malaysia successfully worked with an industrial policy agency, the Palm Oil Research Institute of Malaysia, to effectively break the link between water pollution from crude palm oil processing mills and palm oil production.

These admittedly limited examples of recent "successes" in the East Asian NIEs—along with empirical evidence of substantial variability in some ambient environmental quality indicators (Table 1-2)—suggest that it is time to take another and different look at the evolution of public-sector industrial pollution management policies in the region.[3] That is the purpose of this book, which focuses on case studies of six economies: China, Indonesia, Malaysia, Singapore, Taiwan, and Thailand. This case approach was chosen because it gives a much greater analytical edge through the comparison of a diverse set of studies. The six cases were chosen because they reflect important differences among the East Asian NIEs. These include differences in initial conditions and levels of development, in political institutions and forms of government, in political economies of policymaking, and in the degree to which governments have successfully managed the challenge of pollution during high-speed, urban-based industrial growth.

Because I am interested in redressing what I see as an imbalance in the existing literature, I take a rather unconventional view in these case studies. Rather than focusing on environmental neglect, I look for and document examples of successes in industrial pollution management. But this does not mean that all, or even most, serious environmental problems are being successfully addressed by the East Asian NIEs. Nothing could be further from the truth. As the case studies will show, however, there have been important successes.

Table 1-2. Ambient Air and Water Quality in East Asian Newly Industrializing Economies, 1993

Economy and city	*Ambient air quality*[a]	*Organic water pollution intensity of industry*[b]	*Ambient surface water quality*
China		8.06	38 out of 135 river sections of industrial value had water quality below Grade 5; this limited raw water use to irrigation
Shanghai	246		
Beijing	377		
Indonesia		3.19	Dissolved oxygen and bacteriological quality in water treatment plant in Surabaya were unsafe
Jakarta	271		
Malaysia		1.66	Based on a water quality index; number of clean rivers increased to 38% and number of slightly polluted rivers fell to 64% in all of Malaysia
Kuala Lumpur	85		
Singapore	31	0.42	10 micrograms per liter (µg/l) of biological oxygen demand (BOD) about 68% of the time
South Korea		0.68	n.a.
Seoul	84		
Taiwan		n.a.	Between 1981 and 1999, heavily polluted portions of rivers declined from 14.9% to 12%, while nonpolluted or slightly polluted portions of rivers increased from 46.6% to 66.2%
Taipei	64		
Thailand		1.94	1.5 µg/l of BOD in Chao Phraya (below Thai standard)
Bangkok	223		

Note: In the title, "1993" is an approximation. "n.a." means not available.

[a]Ambient air quality is measured as total suspended particulates or particulate matter, in micrograms per cubic meter.

[b]Organic water pollution intensity of industry is measured as kilograms per $1,000 of industrial value added.

Sources: Except for Singapore and Taiwan, air quality data are from World Bank (2000a, 163). Singapore air quality data are from PCD (1980, 8–9) and water quality data are from PCD (1994, 30). Taiwan air quality data are from "Comparison of Air Quality with Other Countries," July 7, 2000, from the Web page www.epa.gov.tw/english/offices/f/bluesky/bluesky3.htm. Taiwan water quality data are from the Bureau of Water Quality Protection in Bureau of Water Quality, TEPA (2001, 174). China's water quality data are from World Bank (1996a, 91). Water quality data for Indonesia are from World Bank (1994a, 69), and water quality data for Malaysia are from DOE (1994, 18). Organic water pollution intensities of industrial value added are calculated from organic water pollution data in World Bank (2000a, 134–136) and industrial-value-added data in World Bank (1998, 236–237).

The book focuses unabashedly on the political economy of industrial pollution management rather than on the economics of industrial pollution control. This reflects a realization that what needed to be explained was not whether pollution control policies were efficient or cost-effective—it is well known that they are not—but rather why governments chose particular pollution management strategies and how those strategies persisted or changed over time. I also wanted to explain why I found so much variation in pollution management. The more I worked on these issues, the more I realized that the discipline of political economy, rather than that of economics, was most helpful.

The book's focus is also limited to industrial pollution. It does not look at the environment writ large or at urban environmental problems writ large; these would have required consideration of safe drinking water and sanitation, as well as of pollution from households and transportation. Although there is no doubt that these other issues are important (particularly natural resource management issues in Southeast Asia and China and other urban issues almost everywhere), the focus here is on industrial pollution because industry has a privileged place in the political economies of the East Asian NIEs. In each economy, close government–business relationships have taken precedence over the government's relations with other actors in civil society. And each government has erected more or less effective institutional structures to bestow promotional privileges on industry.

Given this history, the book's focus on the political economy of industrial pollution management provides an opportunity to examine how close government–business relationships and the institutions that bestow promotional privileges on industrialists have affected and been affected by the need to address pressing problems of urban industrial pollution. Among other things, the focus also provides an opportunity to ask whether these relationships and institutions have forestalled governments' efforts to clean up pollution, as some have suggested. And it enables me to ask whether governments have been able to take advantage of the institutions of industrial policy to fashion unique approaches to pollution management. This is important because, as is now known, governments have fashioned unique institutional strategies for agricultural and export-led industrial growth. Perhaps they have done the same to control industrial pollution.

The Evolution of Pollution Management Strategies

Because I am most interested in why and how the East Asian NIEs chose particular industrial pollution management policies, rather than whether those policies were efficient or cost-effective, a theoretical framework first used by Haggard (1990) proved to be particularly helpful.[4] His focus was on explaining why the East Asian and Latin American NIEs chose different development strategies and why those strategies persisted and shifted. The focus here is on explaining why six of the East Asian NIEs—China, Indonesia, Malaysia, Singapore, Taiwan, and Thailand—chose different industrial pollution management strategies and why those strategies persisted and changed over time. He cautioned against imputing too much purposiveness and design to economic strategies that emerged at least partly by default, through trial and error and after compromise, and that took years to crystallize.

The case studies that follow in Chapters 2 through 6 suggest that trial and error was an important element of pollution management in each East Asian NIE. But pollution management in each NIE also was subject to compromise and plagued more or less by inconsistency, particularly in China, Indonesia, and Thailand. Haggard argued that strategies, which consist of packages of policies, might usefully be disaggregated because different policies involve different cleavages and conflicts (1990, 23). Much the same can be said for pollution management in the six East Asian NIEs examined here.

Attempts to build comprehensive command-and-control pollution management agencies with the capability to enforce tough emissions standards have not succeeded in all of the East Asian NIEs and have taken time to succeed in others. The case studies below, however, will demonstrate that more focused policies have been quite successful. In explaining economic policy change in the NIEs, Haggard (1990, 28) relied on four distinct, overlapping factors: international market and political pressure, the nature of domestic politics, the structure and institutions of the state, and the development and spread of ideas. As the case studies will demonstrate, the same could well be said about the evolution of industrial pollution management policy in the East Asian NIEs.

International Pressure for a Cleaner Environment

International market pressure—particularly from commodity price shocks and political events such as wars, which sever access to markets—was seen by Haggard (1990, 29–31) as spawning import-substitution industrialization efforts in each East Asian NIE. There is growing evidence that each NIE faces mounting international environmental market pressure, which is provoking policy responses. Sometimes new environmental market access requirements take the form of certification (e.g., the current certification by the International Standards Organization, known as ISO 14000, or eco-labeling of products for sale in countries belonging to the Organisation for Economic Co-operation and Development; Roht-Arriaza 1995). Sometimes they take the form of international environmental supply chain requirements by multinational buying groups (e.g., when the Gap clothing manufacturer and retailer imposes wastewater recycling requirements on its Asian suppliers; interviews in Singapore 1996). Sometimes, environmental pressure simply reflects negative reactions of buying publics and stock markets to bad environmental news of individual firms in the East Asian NIEs (World Bank 2000a, 60–63). Because the East Asian NIEs are so open to trade and investment, it is not surprising that public officials (in ministries of industry, science and technology institutes, and national standards agencies) are worried that this pressure may limit the ability of their economies to export (Rock 1996a). Nor is it surprising that private-sector officials in peak business associations and in leading companies in the East Asian NIEs have similar fears (interviews in Indonesia, Malaysia, Thailand, and Taiwan, 1996).

How have governments and private-sector firms reacted to this new international environmental market pressure? The government of each case-study economy is busy developing ISO 14000 certification and eco-labeling programs (Rock 1996a). In the private sector, peak business associations and leading companies are hard at work creating local business councils for sustainable development and supporting business-oriented environmental nongovernmental organizations (NGOs), such as the Thailand Environment Institute (Rock 1996a). It is too early to tell whether this international environmental market pressure will ultimately affect behavior, or whether efforts by governments and firms to overcome the pressure matter. But there is little doubt that public and private officials are worried that the pressure might matter. At least one recent study (Aden and Rock 1999) confirms that these worries have been translated into investment in pollution abatement expenditures at the manufacturing-plant level.[5]

International environmental market pressure, however, is not viewed in entirely negative terms by policymakers in the East Asian NIEs. They also see it as providing opportunities. In several recent interviews, NIE public officials stated that firms in their economies have learned how to meet the price, quality, and on-time delivery requirements of international buyers. Similarly, they stated that these firms can learn how to meet new environmental requirements more quickly than firms elsewhere (Rock 1996a). This ability might even give the export-oriented firms of the East Asian NIEs at least a short-term competitive advantage. Moreover, public officials in Taiwan have gone a step further by recognizing (before the recent financial and currency crises) that the demand for environmental goods and services in Southeast Asia is likely to grow quite fast during the next quarter-century. Because of this, they are promoting the creation of a domestic environmental goods and services industry, using selective incentives. They have also established quantitative export targets for this new industry, by country and by year. Openness to trade and investment along with late industrialization also creates the possibility that firms in the East Asian NIEs can take advantage of newer and cleaner technologies imported from Japan and the industrial West. There is already some evidence of this in the pulp and paper industry (Wheeler and Martin 1992).

As with economic policy change, international environmental political pressure also appears to be rising and provoking environmental policy responses by the governments of the East Asian NIEs. To prepare for international conferences (e.g., the 1972 United Nations Conference on the Human Environment in Stockholm, 1987 Brundtland Commission, and 1992 United Nations Conference on the Environment and Development in Rio de Janeiro), the governments had to produce position papers outlining the relationship between the environment and development in their economies. The case studies show that the preparation of these papers attracted the attention of political and policy elites and created a political space for those in and out of government who wanted more environmentally friendly pollution management policies. Negotiations over international treaties (e.g., the Montreal Protocol, Convention on International Trade in Endangered Species, and Kyoto Protocol) have required the governments to work out their own positions and to decide whether to sign. Sometimes, economies that are not permitted to be signatories to these treaties (e.g., Taiwan) have taken independent action to demonstrate that they are good environmental citizens in the community of nations. And sometimes, multilateral economic agencies (e.g., Asia-Pacific Economic Cooperation) have forced reluctant governments to reconsider their environmental policies.

As Mathews (1997) so eloquently demonstrates, the growth, spread, and development of international civil society adds to this more formal international environmental political pressure. International NGOs regularly lobby the governments of the East Asian NIEs (along with international organizations working in these economies) to save particular species, to protect forests rich in biodiversity, and to make bilateral and multilateral agreements more environmentally responsible. These governments, like their counterparts in the rest of the world, have also had to contend with the more recent efforts of environmental NGOs in industrial and developing countries to persuade the International Monetary Fund, World Bank, and World Trade Organization to hew to the NGOs' vision of environmental responsibility. Pressure from international aid donors also affects domestic environmental policies by requiring environmental impact assessments for large infrastructure projects, by providing loans for environmental capacity building, and by integrating the environment into all aspects of lending. Occasionally, as in BAPEDAL's PROPER program in Indonesia, donor support can make a critical difference (Afsah and Vincent 2000). Sometimes, international political pressure can more profoundly influence domestic environmental policy. In Taiwan, for instance, the loss of international recognition and pressure from overseas Chinese, particularly in the United States, ultimately helped push a reluctant government to recognize that it might be able to capitalize on significantly improved environmental practices and performance to burnish its international credentials.

Domestic Pressure for a Cleaner Environment

International pressure alone, however, is not likely to bring about economic policy change (Haggard 1990, 43). The same is true for environmental policy change. Much depends on the way governments, domestic firms, and other actors in civil society respond to pressure. What conditions those responses? To begin with, as O'Connor (1994, 5) and others (Hettige et al. 1998; World Bank 1992b) have argued, the relationship between the environment and development (as manifested in rising per capita income) exhibits a consistent pattern, an inverted-U Kuznets curve: As per capita income increases, environmental quality initially worsens; but at some point, this worsening peaks and then declines. (See Figure 1-1 for an example of a Kuznets curve.)

The worsening in environmental quality attending initial increases in per capita income reflects both shifts in the composition of output from less to more pollution-intensive activities (e.g., chemical-intensive agriculture and a shift to more pollution-intensive industries) and a lack of effective environmental regulation. As incomes continue to increase, however, the composition of output begins to shift again, this time to less pollution-intensive activities (services) that reduce the burden of economic activity on the environment. At the same time, governments respond to growing domestic polit-

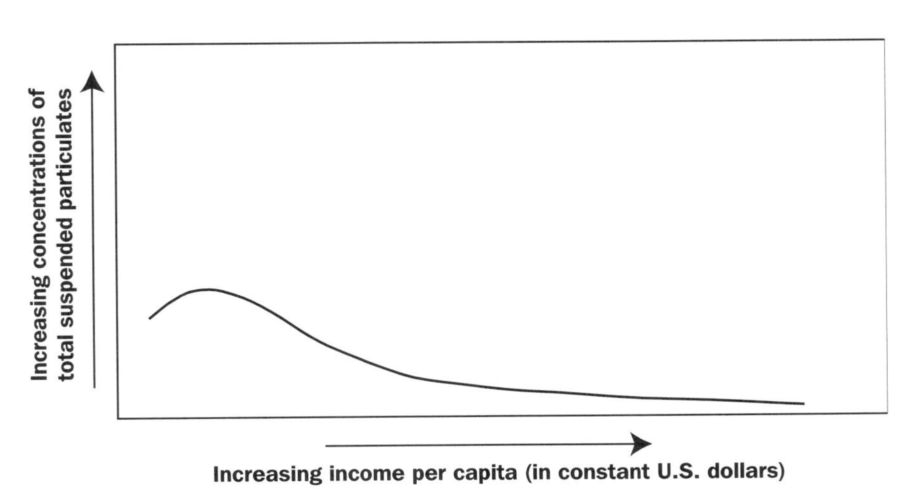

Figure 1-1. Example of an Inverted-U Kuznets Curve

Note: This Kuznets curve shows the change in total suspended particulate concentrations in air as income increases.

ical pressure to clean up the environment by building more or less effective command-and-control environmental regulatory agencies.[6]

The experiences of the East Asian NIEs are largely consistent with this broad pattern. There is little doubt that their early structural shifts in the composition of production contributed to rising portions of inverted-U environmental Kuznets curves for a number of air and water pollutants. For example, during South Korea's post-1965 high-growth era, energy consumption increased two times faster than it did for other upper-middle-income countries, and Thailand experienced a similar development (O'Connor 1994, 25–26). In both instances, this contributed to a decline in ambient air quality. In addition, the rapid growth of the textile sector in Thailand between 1980 and 1989 partly caused an almost equally rapid increase in water pollution. More recently, structural shifts in the NIEs' composition of industry have exacerbated the toxic intensities of production (Brandon and Ramankutty 1993).

These increases in energy and toxic intensities and the shift toward water-pollution-intensive production no doubt have been factors in the declining quality of both air and water. More recent changes in the composition of economic activity in the high-income East Asian NIEs (Malaysia, Singapore, South Korea, and Taiwan) have contributed to less pollution-intensive activities. Thus, after rising between 1964 and 1981, pollution intensities of industrial activity in Taiwan declined, partly because of shifts in the composition of industrial output (O'Connor 1994, 27). Given this broad pattern in the changing composition of output, it is not particularly surprising, as Table 1-2 shows, that ambient air and water quality in the high-income NIEs is noticeably better than in the low-income ones (China, Indonesia, and Thailand).

This inverted-U Kuznets curve tends to be reinforced by the evolution of domestic political forces, which has accompanied high-speed, broadly based economic growth in the East Asian NIEs. Chief among these forces are democratization and the emergence of an educated urban middle class. In the early stages of development—when per capita income is low, most of the population is employed in agriculture, and the incidence of poverty is high—political leaders tend to be more concerned with economic development, poverty alleviation, and food security than with the environment. This lack of official interest in the environment is reinforced by a lack of popular pressure for a cleaner environment and by the authoritarian predilections of governments that retard the development of independent organizations in civil society (Mackie and MacIntyre 1994; Crouch 1996; Girling 1981; Wade 1990; Haggard 1990).

Concern for the environment has increased, however, as per capita income has continued to rise, as education has deepened and spread, and as a greater sense of well-being has taken hold, particularly in the emerging urban middle class. Public opinion surveys now routinely demonstrate that there is no need to convince people that environmental problems are serious (a recent report on Indonesia states this; World Bank 1994a, 177). There is also evidence (particularly from China, South Korea, and Taiwan) that, in response to deteriorating local environments, individuals, unorganized communities, and popular groups in civil society are successfully pressing for local remedies (Cribb 1990; Lee and So 1999). Most often, this has resulted in polluters compensating those affected by their polluting activities.

In most of the East Asian NIEs, the rapid growth so high on the agenda of authoritarian governments contributed to the formation of a large urban middle class, which clamored for more political freedom. In several NIEs (South Korea, Taiwan, Thailand, and now, perhaps, Indonesia), authoritarian governments responded by permitting what appear to be stable transitions to democratic rule. In South Korea and Taiwan, the transition to and consolidation of democratic rule were accompanied by the development of substantial environmental protest movements (Tang and Tang 1997, 1999; Eder 1996; Lee and So 1999).

Although they initially focused on resolving local environmental problems, over time these protest movements developed into national environmental NGOs that built substantial membership bases; undertook studies; published results; lobbied local government officials, legislatures, and executives; and supported political parties and candidates for office (Tang and Tang 1999). Democratization also resulted in a freer press, which reported environmental accidents, lamented the generally poor quality of the environment, and clamored for environmental cleanup. In most NIEs, with Thailand as a notable exception, this combination—along with celebrated pollution incidents—was sufficient to get reluctant governments to enact landmark environmental legislation, establish tougher and tougher ambient and emissions standards, and begin to build competent command-and-control environmental agencies.

Broader support for this pattern can be found in the literature on democracy and the environment (Payne 1995; Congleton 1992). But democratization or the lack of it has not always been a sure guarantee of improved environmental outcomes in the East Asian NIEs. Democratization in Thailand has not been followed either by the creation of a tough and competent command-and-control environmental agency or by improvement in ambient environmental quality. This outcome appears to reflect the elite nature of Thailand's democratic transition and the structure of its democratic institutions. Its current institutional structure favors delay and the underprovision of public goods such as environmental improvement (MacIntyre 2001, 86). Moreover, its elite democratic transition has given business a large voice in policy outcomes and organized groups in civil society a small voice (Karl 1990; Rock 1995). When combined with "money politics," rampant vote buying in the Thai countryside, and weak and short-lived coalition governments, it is not surprising that Thai democratization has not been accompanied by environmental improvement.

This outcome in Thailand is also consistent with at least some of the literature suggesting that democracies might not be particularly hospitable to better environmental outcomes (Achterberg 1996; Fiorino 1989). A more democratic Philippines with an active and vociferous NGO movement has been similarly slow to respond to a deteriorating environment. Conversely, less democratic Malaysia has been able to effectively sever the link between pollution from crude palm oil mills and the production and exporting of palm oil (Vincent et al. 2000). Distinctly authoritarian Indonesia has also been credited with significant innovation in pollution management (World Bank 2000a; Aden and Rock 1999). Similarly, Singapore's nominally democratic but authoritarian government built an effective command-and-control environmental agency during the early stages of industrialization. Today, the ambient environmental quality in Singapore is similar to that in the countries that belong to the Organisation for Economic Co-operation and Development (OECD). This variation suggests that environmental outcomes may well depend less on the type of regime than on the nature of the state, its institutional structure, and the relationship of each to civil society.

The Role of the State in Cleaning Up the Environment

There is little doubt that the actions of states loom large in industrial pollution management. The task—correcting market failures or internalizing externalities—is one that naturally falls to the public sector. Because of this, it is not surprising that virtually all of the East Asian NIEs, including all those with case studies in this book except Taiwan, launched their environmental improvement strategies by passing landmark legislation. Where governments in these economies varied, however, was in their ability to build effective traditional public-sector command-and-control environmental agencies with the legal authority and tools to monitor and enforce emissions and ambient standards.[7] O'Connor (1994) and World Bank researchers (1992a, 1993, 1994a, 1994b, 1997b) describe this process in some detail for each East Asian NIE. Each step took time, and in several instances they were propelled forward by notorious pollution incidents. This is very similar to what had happened within the OECD countries a generation earlier (Lovei and Weiss 1997).

In the East Asian NIEs, however, government strategies for environmental improvement went well beyond the creation of traditional public-sector command-and-control environmental agencies. In Singapore, the creation of a tough command-and-control agency went hand in hand with the integration of the environmental actions of a new Ministry of the Environment into the promotional activities of the Economic Development Board and with the industrial siting and infrastructure activities of the Jurong Town Corporation. In this way, Singapore effectively embedded environmental considerations in the broader institutions of industrial policymaking. Just the opposite happened in Taiwan, where the government deliberately bypassed important industrial policy agencies (e.g., the Industrial Development Bureau in the Ministry of Economic Affairs) while creating a tough command-and-control agency and granting it sufficient authority to effectively monitor and enforce emissions standards. As part of this process, government officials built new relationships with old and new actors in civil society (academic and other environmental experts, communities, and NGOs) over a new issue: cleaning up the environment. This forced several of Taiwan's premier institutions of industrial policy, including the Industrial Development Board and the Industrial Technology Research Institute, to develop their own unique industrial environmental improvement strategies.

In China and Indonesia, effective environmental responses to industrial pollution were less dependent on creating landmark legislation and comprehensive command-and-control environmental agencies. Because of this, pollution management successes in these economies were more limited: cutting wastewater emissions from large plants along major rivers in Indonesia, getting manufacturing plants in one city on Java to begin to invest in pollution control, and improving environmental quality in some of China's largest cities. In both economies, limited successes relied on information-based strategies that depended on the ability of managers in relatively weak environmental agencies to craft innovative solutions to particular problems that also attracted the attention and support of political leaders or those in more powerful economic agencies. Finally, in Thailand, effective pollution management and even limited, targeted pollution control successes have proven elusive.

Given these differences in governmental response to accumulating industrial pollution, it becomes important to ask why some East Asian NIEs appear to be so much better at pollution management than others, and why state responses look so different. As was indicated above, some of the differences reflect variances not only in the domestic politics of pollution management but also in the state's capacity to attack rising pollution loads. Haggard (1990, 44–45) argues that three characteristics of the state bear on its ability to effect change in economic policy, and his point applies to environmental policy as well. First, state actors must have some degree of insulation or autonomy from pressure groups in civil society that might impede the ability of the state to enact politically sensitive policies. Because change in both economic and environmental policy is likely to be politically sensitive to and opposed by business interests that profit from the status quo, some degree of autonomy from businesses is likely to be particularly important for successful pollution management. Second, effective state action is likely to be easier when the state has a cohesive, technocratic, pragmatic, and goal-directed bureaucracy. When the bureaucracy lacks these characteristics, governmental decisionmaking is likely to be subject to patron–client ties or too much interest-group pressure. This makes effective, concerted state action more difficult. Third, effective state action is likely to be facilitated or constrained by the range of policy instruments open to policymakers. If the state is able to innovate to expand the range of policy instruments, they are likely to be more successful.[8]

To these three characteristics of states—autonomy from those who are most likely to be opposed to environmental improvement, cohesiveness of state decisionmaking structures, and the range of policy instruments available to state managers—two others need to be added. First, state actors are likely to need what Evans (1995, 12–13) labels "embedded autonomy" or "autonomy for independent action," which is rooted in institutionalized channels of communication between state actors and those in the private sector. These channels allow both the freer flow of information that policymakers need for effective regulatory policies and the building of trust that makes environmental cleanup possible at a reasonable cost.

Second, states are likely to need a structure of political institutions—electoral systems, institutional divisions of governmental power, and party systems—that strengthens their ability to enact, implement, and sustain commitment to new policies (Cox and McCubbins 2001, 24–27). But as MacIntyre (2001, 85) argues, the flexibility needed to enact and implement new policies pulls in a different direction from the stability needed to sustain commitment to those policies.[9] The commitment to sustaining new policies is likely to be greatest if control over policy is spread across a number of actors or potential vetoers who can rein in those trying to undermine a particular policy. Conversely, implementation of new policies is likely to be greatest in an institutional framework where control over policy is concentrated. As recent research has demonstrated (MacIntyre 2001; Haggard and McCubbins 2001), how governments are organized has important implications for the ability of governments to enact, implement, and sustain commitment to new policies.

How do the governments of the East Asian NIEs compare along these dimensions? The governments of Singapore, Taiwan, and to a lesser degree Malaysia lie at one end of the spectrum. Each of these governments tends to have substantial autonomy from business interests and relatively cohesive, technocratic, and pragmatic decisionmaking structures. Each also possesses a demonstrated ability to develop innovative policy instruments to tackle pressing development problems, institutionalized channels of communication between state actors and the private sector, and institutional structures that favor policy flexibility and policy stability. It is not surprising that the case studies reveal that each government has used these advantages to create tough and competent, but fair, command-and-control environmental agencies that are embedded in its unique institutions of industrial policy. But the case studies also reveal that each has done so in quite different ways.

The government of Thailand is at the other end of the spectrum (Rock 2000b). Since democratization, the Thai government has been deeply penetrated by business interests. This has severely compromised the state's autonomy for independent policy action, particularly actions that do not serve business interests. Moreover, the multiple-party coalitions ruling Thailand's parliamentary democracy have undermined the limited cohesiveness of the state's decisionmaking structures (as have money politics, rampant vote buying, and the division of spoils that follows each election cycle), particularly the structures of the core economic agencies that existed before democratization. This means that technocrats in such state organizations as the National Economic and Social Development Board—who might have led an environmental improvement strategy from within the bureaucracy—have been undercut.

Moreover, because democratic Thai governments have been based on weak, short-lived coalitions, policymakers in the country's environmental agency have not been particularly adept at expanding their range of policy instruments. To make matters worse, as Hicken (1998, 1999) and MacIntyre (2001, 90–94) have shown, the structures of Thai democratic institutions favor delay, underprovision of public goods, and zero-sum bargaining by a substantial number of potential vetoers. The state has neither the policy decisiveness and flexibility to enact new policies nor the policy resoluteness and stability to sustain them. Thus it is not particularly surprising that Thailand, unlike the rest of the East Asian NIEs, has not had even one moderate success in controlling industrial pollution.

The governments of China and Indonesia fall between these polar opposites. In Indonesia, as in Thailand, pervasive patron–client ties between government and business support rent-seeking behavior (i.e., the use of bribes, coercion, and lobbying to increase one's income) and appear to circumscribe the government's policymaking autonomy (MacIntyre 1994). Because of these patron–client ties and consensual decisionmaking styles, decisionmaking structures in Indonesia also appear less coherent, technocratic, and pragmatic; and they are more open to manipulation by particular business interests than those in Singapore and Taiwan. They also appear to be more subject to veto by powerful business interests. One might expect this combination, as in Thailand, to have blocked the implementation of policies to curb industrial pollution. Yet some improvement is noticeable. The question is, Why?

The detailed case study of Indonesia in Chapter 4 will demonstrate that improved outcomes—at least in the polluting behavior of the largest factories along the country's major rivers and in plants in one of the larger cities on Java—flow from the innovative actions of environmental policymakers within the interstices of the state. It is interesting that these actors used their strong ties to President Soeharto (alternatively spelled "Suharto") to persuade him to support several innovative pollution control programs (interviews with senior officials of BAPEDAL, 1996). They also used technical assistance and their relationship with the World Bank to shield them from criticism. And they were exceedingly pragmatic and sensitive to potential political opposition in their design of pollution control policies. In this instance, as in many others, state policy ended up being more coherent than conventional wisdom suggests (Rock 1999). As Liddle (1991) has argued, this is because President Soeharto often looked beyond narrow patron–client interests to support important policy changes. In these instances, he did so by sanctioning the use of public disclosure of the environmental performance of major water polluters and by allowing the mayor of a large city to start his own pollution control program. Thus, he signaled to business interests that they had better start addressing the pollution associated with their manufacturing plants. At the same time, because the government had not reached consensus regarding the environment, the president was not willing to sanction the development of a more effective, comprehensive command-and-control environmental agency. This meant that a number of donor-funded institution- and capacity-building projects in BAPEDAL met with extremely limited success.

The government of China exhibits some of the same characteristics as that of Indonesia. It does not possess much autonomy from business interests. It is less cohesive, if not less technocratic and pragmatic. And its institutions and political economy of policymaking favor delay. Despite this combination, a relatively weak environmental agency in China devised and implemented a unique city-level environmental rating, ranking, and public disclosure program that ultimately captured the attention of political elites (mayors) in the country's major cities. Once mayors started asking why their city ranked lower than others, how their city could improve its ranking, and what this might cost, city environmental managers started working with powerful city economic agencies to answer these questions. By doing this, they learned how to take advantage of China's bargaining model of policy implementation to overcome the stalling actions of a large number of potential vetoers (Walder 1992; Lampton 1992). And they linked these efforts to an environmental target system that required mayors and provincial governors to sign environmental responsibility contracts with those immediately above them in China's bureaucratic hierarchy. Over time, this combination proved effective in fostering significant improvements in environmental quality.

The Role of Ideas

Without what Haggard (1990, 45) refers to as a more or less coherent framework of policy-relevant knowledge (or ideas), it is doubtful that international and domestic pressure to clean up the environment would have been sufficient to get the East Asian NIEs to act. Nor would the NIEs with more coherent decisionmaking structures, more embedded autonomy, and better institutional balance between decisiveness and resoluteness have been likely to improve environmental outcomes. This suggests that a shift in ideas regarding the relationship between the environment and development (economic growth) probably played some role in the NIEs' movements toward a cleaner environment.

For most of human history and for much of the world, including the East Asian NIEs, the dominant idea regarding the relationship between growth and the environment is embodied in the phrase "grow first, clean up later." This phrase reflects the notion that neither poor people nor poor (developing) economies can afford to invest in the environment. Economists have given credence to this view with talk about a clean environment being a luxury "good." There is little doubt that, until recently, political leaders and policymakers in developing countries and international organizations thought and acted this way. Thus it is not surprising that most developing economies, including the East Asian NIEs, have followed the "grow first, clean up later" environmental strategies adopted a generation ago by the countries belonging to the OECD. Nor is it surprising that most donor organizations, including the World Bank, have been so slow to respond to the concerns of the environmental community (Wade 1999).

Yet it is also true that there have been significant shifts in what can only be seen as a coherent framework of policy-relevant knowledge regarding economic growth and the environment. The Brundtland Commission's definition of sustainable development struck a responsive chord with economists, policymakers, international organizations, and aid donors. As a result, all of these groups now think about development in Brundtland terms (World Commission on Environment and Development 1987). Empirical research—on environmental Kuznets curves (Hettige et al. 1998; World Bank 1992b), on the human health costs of environmental degradation, on cost-effective industrial pollution management policies (World Bank 2000a), and on potential win–win opportunities in pollution prevention and clean production (Angel and Rock 2000)—has reinforced this trend. Thus there is now substantial evidence in each East Asian NIE that political leaders and economic policymakers, who once complained that environmental agencies might slow growth, now complain that these agencies are not doing enough.

The government of each East Asian NIE also has responded to the range of new policy ideas about how best to control industrial pollution. Given the evolution of pollution control policies in the OECD—from command-and-control to market-based instruments and information-based strategies—it is not particularly surprising that East Asian NIEs began their strategies by investing in the creation of very traditional command-and-control agencies.

Nor is it surprising that several of these agencies continue to rely heavily on prescribed technology-based strategies or that several others have tried to take advantage of market-based instruments and information-based policies. In China and Indonesia, where policymakers have not been totally successful in creating such agencies, innovative actors in nascent agencies have relied particularly on information-based policies and market-based instruments.

Toward a Political Economy of Environmental Improvement

Because each East Asian NIE is relatively open to foreign trade and investment, both government and the private sector have been subject to growing international economic and political pressure to clean up the environment. Moreover, a relatively free flow of policy-relevant ideas within the political economy of each NIE has enabled political leaders, policymakers, and researchers in universities and think tanks to become familiar with the shifts in ideas regarding the relationship between development and the environment. Both of these situations have made it easier for NIE governments to embark on industrial environmental improvement strategies. But neither, by itself, has been sufficient to get the governments to address rising industrial pollution. The evidence in the following chapters will demonstrate that the variation in the domestic politics of pollution management and in the capability of governments regarding management programs accounts for differences in both the strategy employed and outcomes.

In Singapore, the prime minister, Lee Kuan Yew, a near-benevolent despot committed to reform—along with a strong, autonomous state with a coherent, technocratic, pragmatic decisionmaking structure and institutionalized channels of communication with the private sector—helped to launch the earliest, most successful industrial pollution management program in East Asia. This enabled Singapore to "grow while cleaning up" its environment. In 1970, the prime minister communicated his concern about the environment to the public and the bureaucracy by creating an antipollution unit in his office. In 1972, he created a traditional command-and-control Ministry of the Environment (ENV). He then left it to the bureaucracy to figure out how to successfully integrate environmental considerations into the broader institutions of industrial policy.

The pragmatism of the new ENV's senior officials proved particularly important. These individuals knew, particularly in the early days, that they could not be too hard on polluters. They also knew that they had to win the confidence (overcome the veto powers) of their counterparts in the Economic Development Board (EDB), which granted investment promotional privileges, and in the Jurong Town Corporation (JTC), which sited promoted firms and provided them with infrastructure. And they knew that they were expected to make a difference. They resolved problems pragmatically. They traveled abroad to identify international best practices for pollution control. They compiled and distributed lists of suppliers of control equipment to firms. They set emissions standards on the basis of international best-practice technologies, and they gave polluters ample time to clean up. Most important, they devised operating procedures that involved both the EDB and JTC in decisionmaking.

Semidemocratic Malaysia, like Singapore, also found a way to "clean up while growing" rapidly. But unlike Singapore, the pressure to clean up came from outside the government and from an important voting constituency: rural Malays, who expressed increasing dissatisfaction over rising pollution loads from the numerous crude palm oil (CPO) processing mills that dotted the Malay countryside. Because the dominant party in government, the United Malay National Organization, depended on votes from rural Malays for its control of the government, it could not easily ignore this constituency or the inroads made into it by an important Islamic opposition party. This combination forced the government to draw on its considerable state strength and embedded autonomy, as in Singapore, to empower a traditional command-and-control environmental agency to clean up CPO pollution. But, as in Singapore, it did more than just impose tough emissions standards on CPO polluters. It used its embedded autonomy to synchronize its strengthening of emissions standards with the introduction of best-practice control technologies. This made it possible to almost totally unlink CPO production and exports from CPO emissions.

The Taiwanese government, particularly before democratization, was relatively autonomous and characterized by a coherent, technocratic, and pragmatic decisionmaking structure and by institutionalized channels of communication with the private sector. However, unlike Malaysia and Singapore, it followed a "grow first, clean up later" environmental strategy. Its early environmental neglect appears to have been the consequence of its tight links with businesses, which led it to focus on growth and exports at the expense of nearly everything else (Chun-Chieh 1994). After democratization, the policymaking environment was quite different from that in Singapore. With political liberalization, the government of Taiwan faced a hostile press, a growing environmental protest movement, and opposition political parties anxious to blame the government and the Kuomintang (KMT) for the country's deteriorating environment (Tang and Tang 1997).

Yet within the Taiwanese government, a strong, autonomous industrial policy agency—the Industrial Development Bureau of the Ministry of Economic Affairs—used its veto power to oppose environmental cleanup. It did so because it feared this would slow growth at a time when industry was being "hollowed out" by rising wage rates and an appreciating currency. At this point, the leadership of the KMT appeared to be trapped between escalating public demands to reduce pollution and a growing reluctance to do so in the interstices of the state's economic agencies. The KMT's senior leaders appear to have concluded that their ability to compete in future elections in democratic Taiwan depended, at least in part, on some cleaning up. Given the Industrial Development Bureau's unwillingness to comply, the government responded by completely bypassing the bureau in creating a powerful command-and-control environmental agency. Because of this, neither the Industrial Development Bureau nor representatives of industry were permitted to participate in the many expert panels used by the Taiwan Environmental Protection Administration to set ambient and emissions standards. This ultimately forced each agency to develop its own environmental improvement strategies.

In both China and Indonesia, the state had less autonomy from business, and decisionmaking structures were riddled with patron–client ties and rent-seeking. In addition, where institutionalized opportunities for the exercise of veto power were substantial, innovative actors in relatively weak environmental agencies developed partial, incomplete, narrowly targeted approaches to pollution control that drew on information-based policies and market-based incentives. In Indonesia, the central government relied on an innovative public disclosure campaign to get large factories along the country's major rivers to abate their water pollution, whereas at the local level, the mayor of a large city initiated his own limited environmental monitoring and enforcement program. Both occurred within the context of an otherwise moribund pollution management agency, and both required the approval of the country's president. For its part, a weak environmental agency in China devised a unique and effective city-level environmental rating, ranking, and public disclosure program that ultimately attracted the attention of city mayors.

These differences suggest multiple political pathways to improved environmental outcomes or (in Thailand) blocked pathways to improved outcomes. The substantial, sustained commitment of such political elites as Lee Kuan Yew in Singapore to a cleaner environment—in combination with a coherent, technically competent bureaucracy with substantial embedded autonomy—provided a top-down political pathway to improved outcomes. The growth of the urban middle class and transitions to democratic rule in several East Asian NIEs, most notably South Korea and Taiwan, provided a second political pathway to improved outcomes (Tang and Tang 1997, 1999; Eder 1996). This suggests that even old authoritarian governments and parties can become engaged in cleaning up, if they see cleanup as in their vital interest. The response of semidemocratic Malaysia to a rising crescendo of protests from an important political constituency provided a third pathway to improvement.

Despite differences in their initial conditions, level of development, and politics, Malaysia, Singapore, and Taiwan all took advantage of their strong, autonomous states with substantial embedded autonomy in dealing with the private sector. This autonomy enhanced the policy flexibility of these economies. Embedded autonomy made it possible to draw on the trust gained by years of positive collaboration between government and the private sector so that reductions in emissions and improvements in ambient environmental quality did not threaten profitability or exports. This made it easier to sustain commitment to industrial environmental improvement. This finding suggests, contrary to what some have argued, that these governments can be as successful with regulatory policies as they have been with promotional policies.

It is also important to note that the creation of tough, competent, pragmatic, and fair command-and-control environmental agencies with sufficient capacity and legal authority to monitor and enforce new emissions standards was the sine qua non of success in each of these economies. For the most part, these agencies were modeled on their counterparts in the OECD countries, particularly the U.S. Environmental Protection Agency. This suggests that there may be fewer political, social, and cultural impediments to such transfers than was thought. Where pollution control strategies differed in these economies is in how the new agencies interacted with both the public and key industrial policy agencies.

For those states with less embedded autonomy, less sustained commitment from political elites, and decisionmaking structures subject to rent-seeking, a pathway for limited environmental cleanup was open to policy innovators who devised targeted solutions to environmental problems. The success of these solutions in both China and Indonesia depended on a range of factors. A minimum level of capacity in environmental agencies was necessary for innovation. The ability of those in environmental agencies to attract attention and continuing support from powerful political leaders was also important. Beyond that, successful innovations started small and focused on pressing problems. The designers and implementers of these small, focused, innovative policies were also quite good at anticipating and co-opting potential political opposition. These factors illustrate how even weak environmental agencies in "soft" states can get some industrial polluters to abate their emissions. But when these factors were either missing or combined with a structure of political institutions that rewarded zero-sum political bargaining, as in Thailand, multiple vetoers forestalled effective government action.

Notes

[1]Through the remainder of this book, the economy of "Taiwan, China" will be referred to as "Taiwan."

[2]It is important to note, however, that the improvements cited in Hettige et al. (1996) are not related to the effectiveness of governments' industrial pollution control programs.

[3]Extrapolating from Table 1-2, it can be seen that Singapore and Taiwan have achieved Organisation for Economic Co-operation and Development (OECD) levels of urban ambient air quality; South Korea and Malaysia are not far behind; and urban ambient air quality in China, Indonesia, and Thailand is among the dirtiest in the world. Although surface water quality data are not directly comparable, it appears that a similar pattern emerges for ambient surface water quality and organic water pollution from industry. Singapore, Malaysia, and Taiwan have less polluted rivers and lower organic pollution intensities per $1,000 of industrial value added than China, Indonesia, and Thailand.

[4]Chapter 7 does consider whether the end-of-pipe policies that predominate in the region are the most cost-effective. The analysis in the chapter suggests that they are not.

[5]Aden and Rock (1999, 367) show that environmental market pressure affects the plant-level pollution abatement expenditures in a sample of manufacturing plants in Indonesia.

[6]By command-and-control environmental agencies, I mean ones that tend to rely on technology-based emissions standards rather than market-based incentives or information-based policies to get polluters to clean up. As Sterner (forthcoming, 65) argues, this is how environmental agencies in the OECD began to clean up industrial pollution. As he says, this preference for technology-based standards lies in the fact that it simplifies both what regulators regulate and what polluters have to do.

[7]As in the OECD, preference for regulation of technologies in environmental agencies in the East Asian NIEs reflects the fact that it simplifies the tasks of both regulators and the regulated. Therefore it is not surprising, as all the case studies except for China show, that sustained success in reducing emissions and improving ambient environmental quality has depended on the creation of credible command-and-control agencies.

[8]It is important, however, to note that the range of policy instruments available to state actors may well be endogenous. That is, an autonomous state with a pragmatic, goal-directed bureaucracy may be more effective at expanding the range of policy instruments than those that are less autonomous and less goal-directed.

[9]To be more precise, MacIntyre (2001, 85) refers to policy decisiveness as policy flexibility and policy resoluteness as policy stability.

2

Integrating Pollution Management into Industrial Policymaking in Singapore

Singapore is the only East Asian newly industrializing economy that did not follow a "grow first, clean up later" environmental strategy. In 1969, the government launched its all-out strategy to leapfrog to industrial-world status by attracting foreign multinationals and creating an environmentally responsible Singapore with industrial-world health, education, telecommunications, and transport facilities (Lee 2000, 56–57).[1] Its development of an industrial pollution management strategy was not far behind. The very next year, the prime minister, Lee Kuan Yew, created an Anti-Pollution Unit (APU) in his office. He created this unit so that the environmental impact of new investment attracted by the Economic Development Board (EDB) could be evaluated before the offering of promotional privileges (interviews with Ministry of the Environment officials, winter 1995–1996).

By this act, the prime minister signaled to the country's powerful economic development agencies that economic growth in Singapore would not come at too great an expense to the environment. In 1972, the government created a Ministry of the Environment (ENV). Clean Air Standards Regulations were promulgated in 1972, and the Clean Air Act was enacted in 1975. A Water Pollution Control and Drainage Act was passed in 1975, and Trade Effluents Regulations covering numerous water pollutants were announced in 1976. By 1979, Singapore was well on its way to growing a rather clean economy.[2] By the early 1990s, ambient air and water quality in Singapore was equal to that in the countries that belong to the Organisation for Economic Co-operation and Development (OECD).[3]

Unlike in South Korea or Taiwan, environmental improvement in Singapore was not led or pushed forward by international pressure, by pressure from organized groups in civil society, or by widely publicized pollution incidents.[4] The government has long made a habit of ignoring international pressure;[5] organized groups in civil society exert little, if any, influence on either economic or environmental policy (Kirtada and Heike 1995, 293–298); and Singapore has not had even one substantial pollution incident (Lin 1997, 246). Unlike in China, Indonesia, Malaysia, and Thailand, patron–client ties between government and business in Singapore did not forestall environmental improvement. Because improvement in Singapore began so early, more recent ideas about sustainable development and the high costs of environmental degradation, which proved so important to launching improvement programs in the rest of the East Asian newly industrializing economies, exerted little influence on Singapore's strategy.

What drove environmental improvement in Singapore? There is only one possible answer to this question: the government. More precisely, as will be demonstrated, environmental improvement in Singapore was led by the substantial and sustained commitment of a near-benevolent despot: Prime Minister Lee. And it was driven by the decisions of pragmatic technocrats in the ENV who were embedded in a coherent industrial-policy decisionmaking structure within Singapore's strong, autonomous state. This suggests that Singapore was successful, from an early date, in integrating environmental considerations into the machinery of both economic and industrial policymaking.

Why and how did this happen, however? As of yet, there are no good answers to these questions. Several studies (Ling 1995; Foo et al. 1995; Koe and Aziz 1995; Tay 1993) describe the evolution of environmental law, environmental regulation, and industrial pollution management. Others have demonstrated that ambient air and water quality in Singapore rivaled that in the OECD countries from a fairly early date (PCD 1980, 1994). But none of these studies explains either how Singapore grew so fast for so long without experiencing a significant deterioration in ambient air and water quality or how the country improved ambient air and water quality despite a massive expansion of manufactures and manufactured exports.[6] Except possibly for Lin (1997), who describes how environmental considerations affected the siting of industrial plants, none of these studies examines how or whether management of industrial pollution was integrated into the institutions of industrial policy that have been credited with so much of Singapore's developmental success.[7]

Development of a Strong, Autonomous State

Singapore's developmental and environmental successes did not come easy, however. Despite the fact that it possessed several early advantages over its neighbors in Southeast Asia, from the perspective of the early 1960s, the future of its economic development and environmental improvement was by no means guaranteed.[8] At that time, the economy was dominated by a declining reexport trade in staples, the population was growing faster than employment opportunities, and manufacturing consisted of an extremely limited number of small family-based enterprises. The collapse of its federation with Malaysia in 1965 left Singapore in an extremely difficult position (Haggard 1990, 102, 105). Economic difficulties stemming from declining trade with Malaysia were exacerbated by Indonesia's confrontation with Singapore and Malaysia (which was mostly saber rattling, with limited armed conflict) and by a planned withdrawal of the British military from Singapore. The former undermined what was left of Singapore's trade with its neighbors, whereas the latter eventually cost Singapore about 70,000 jobs and 20% of its gross domestic product (Lee 2000, 52). At this point, the country suffered from a high-wage economy, rising unemployment, and limited opportunities for industrialization. Under these circumstances, little attention was paid to the environment, which was subject to smoky and dirty air from wood-processing activities and open burning and dirty water from pig raising, dumping of raw sewage, and textile dyeing (interviews with ENV officials, 1995–1996).

To make matters worse, internal political turmoil made it difficult to consistently pursue any development or environmental strategy. Politically "moderate" leaders within what was to become the major political party, the People's Action Party (PAP), were contending for control of the party and government with leftist elements within the PAP and a leftist labor party (Haggard 1990, 104–107, 112). By 1963, moderate elements within the PAP had finally wrested control of the party away from the left by politically outmaneuvering it and by relying on repression, arbitrary arrests, and detentions under an Internal Security Act. They subsequently turned their attention to gaining control of the government. This was accomplished by taking over the City Council and by asking voters for a "sweeping electoral mandate" for "strong and decisive rule," which could be used to develop a program that cultivated economic ties with the rest of the world, particularly the West (Means 1998, 98). The PAP's leaders saw this as essential to Singapore's economic and political survival. This mandate was granted in the 1968 elections, when the PAP captured all 58 parliamentary seats. Singapore subsequently became an essentially one-party state. Opposition political parties rarely won more than a few seats in the legislature.[9] When they did, the government did not shirk from using extraordinary means to deny even these few seats to the opposition.[10]

The cementing of political control over party politics and government was followed by an attack on Singapore's high labor costs.[11] This was facilitated by earlier efforts that had successfully depoliticized trade unions and replaced them with a government-controlled National Trade Union Congress (Huff 1995, 1424). Now, high labor costs were attacked by an Employment Act that lengthened the work week; reduced the number of holidays; and placed restrictions on retirement benefits, paid leave, overtime, and bonuses. An Industrial Relations Act reinforced the Employment Act by limiting the ability of unions to represent managerial employees; exempting promotions, transfers, and firings from collective bargaining agreements; and lengthening labor contracts (Haggard 1990, 112). After 1972, wage growth was successfully constrained by the tripartite National Wages Council.[12]

The politically moderate, English-educated leadership of the PAP also made sure that Singapore's rather weak Chinese business community remained small, weak, and marginalized. (What follows in this and the next paragraph, unless noted, is drawn from Huff 1995, 1431–1432.) Opposition to the Chinese business community within the PAP partially reflected the community's basic antipathy toward government. It also reflected the community's support for China, Chinese culture, and Chinese language rights. Prime Minister Lee signaled his opposition to the goals of the Chinese business community by promoting English as the language of school instruction and by attacking a key spokesperson for this community, Tan Lark Sye, and the Chinese-language Nanyang University that he founded. Both were seen by many in the Chinese business community as bastions of Chinese culture.

Yet Lee condemned Tan, the university, and its governing council in unduly harsh terms as little more than a front for the Chinese Communist Party (Huff 1995, 1432). He was ultimately successful in seeing the university merged with the English-speaking National University of Singapore. It is not surprising that the government of Singapore never seriously considered adopting a Hong Kong–like model of development based on a large number of Chinese-owned small- and medium-sized manufacturing enterprises. Because of this, businesspeople in Singapore neither decide policy "nor ... exert pressure on the government" (S.Y. Lee 1978, 50).

The PAP also learned how to control the media. To begin with, Singapore's constitution does not guarantee a free press (Means 1998, 102–103). The government also seems to think that the mass media should play an important role in the country's development (Lim 1983, 96). Thus it is not surprising that all five radio stations and all three television stations in Singapore are owned and operated by the state-owned Singapore Broadcasting Corporation. Singaporeans have been routinely denied access to satellite dishes unless economic need for them is demonstrated; and as late as 1995, few if any Singaporeans had access to the highly popular Asian Star-TV beamed out of Hong Kong (interviews in Singapore 2001). Government control over print media is equally strong. Although newspapers are privately owned, the government influences the appointment of their editors and boards of directors, and the sale of shares of individual newspapers.

The government can and has limited the circulation of publications that it deems guilty of "distorted reporting" (Means 1998, 103). In the past, the government has penalized the *Far Eastern Economic Review,* the *Asian Wall Street Journal, Time* magazine, *Asiaweek,* and the *International Herald Tribune* for such reporting (Means 1998, 103). The government has also filed libel suits against its critics in the media, and on several occasions the courts have ruled in the government's favor. The most important consequence of Singapore's approach to the media is that it has muted criticism of both the government and government officials. This includes reporting on the environment, which is largely limited to "reinforcing government policies."[13]

Tight government control over opposition political parties, organized labor, Chinese capital, and the media extends to other organized groups in civil society. With few exceptions, such groups simply do not exist in Singapore. This is not to say that there are not organized channels for policy input, but these channels are tightly controlled and largely hidden from public view.[14] As a consequence, there is little public debate over policy. And thus there is no evidence in Singapore of the widespread practice in the rest of Asia whereby neighbors, communities, and environmental nongovernmental organizations (NGOs) exert public pressure on polluters and on government to get polluters to reduce emissions (Kirtada and Heike 1995, 296).[15] The government has also been known to use its access to resources to keep voters in line. Voting constituencies have been openly manipulated to manage ethnic tensions (Means 1998, 100), reward electoral supporters, and disperse political opposition (Huff 1995, 1432).

The net effect of these actions has been the creation of a strong, autonomous state and weak organizations in civil society. Within this state, economic policymaking has been limited to a small, honest, well-trained, well-paid, and highly capable and pragmatic technocratic elite concentrated in several new public-sector institutions, such as the EDB, the Jurong Town Corporation (JTC), and the Development Bank of Singapore (DBS). Initially, these institutions promoted import-substitution industries such as car assembly, television and radio manufacturing, and shipbuilding. But once these efforts started to flounder and Indonesia and Malaysia turned against Singapore, senior government officials, including the prime minister, decided that Singapore's salvation lay in leapfrogging to industrial-world standards, much as Israel had done in the Middle East (Lee 2000, 49–52, 57).

Development of Industrial Policy Institutions

As the former prime minister stated in a recent book on Singapore's development, there were two elements to the country's strategy. First, Singapore would link itself to the West by attracting multinational corporations, particularly those based in the United States. Second, the government of Singapore would create an industrial-world oasis, with high-quality public and personal safety and security, and health, education, telecommunications, transport, and services (Lee 2000, 56). Once the strategy had been devised, the government turned to creating the institutions and policies to manage Singapore's export-oriented industrial transformation, which relied heavily on foreign investment and multinational corporations (Huff 1995, 1425–1426).[16] The implementation of promotional privileges for investors—particularly foreign investors—promulgated under the various Pioneer Industries Acts, was left to the EDB (Times Academic Press 1993).

Initial Steps

Initially, the EDB was responsible for identifying industries to attract to Singapore; providing them with good infrastructure in industrial estates, including factories with rentable space, sewer lines, roads, and electricity; and providing subsidized credit (interviews with ENV and EDB officials, winter 1995–1996). Over time, some of these tasks were taken over by other agencies. By the early 1970s, the major public-sector institutions that guided Singapore's export-oriented industrial transformation were in place. The EDB scoured the international environment for new industries to attract to Singapore and offered them lucrative promotional incentives. The JTC assumed responsibility, formally vested in the EDB, for physical infrastructure development, including roads, sewer lines, electricity, and flatted factories in industrial estates. The DBS took over the EDB's responsibilities for subsidizing loans for promoted industries, particularly for state-owned enterprises and government-linked companies. The International Trading Company assumed responsibility for export promotion (Rodan 1989, 94).

Two of these agencies—the EDB and JTC—proved particularly important to Singapore's ability to attract foreign investors. The market-friendly EDB developed criteria for investment promotion that addressed Singapore's development challenges. In the 1960s and early 1970s, promotional privileges were used to attract labor-intensive industries to address the country's underemployment problem. Once full employment was achieved, the EDB's promotional privileges were used to attract industries with higher value added and that were more skill intensive. More recently, promotional privileges have been used to attract multinational corporations willing to make Singapore their regional operating centers for Asia or expand their research and development activities in Singapore (interviews with EDB officials, winter 1995–1996). The promotional activities of the EDB are pragmatic and subject to substantial administrative discretion. Its officers spend substantial time and energy working with clients to understand their objectives. If those objectives are consistent with where Singapore wants to go, the EDB can and does sweeten promotional privileges. This suggests that the EDB profits from a degree of what Evans (1995, 12–13) labels "embedded autonomy," but neither access to promotional privileges nor levels of privileges are automatic. Because of this, Singapore does not have a "one-stop shop" for investment licensing.

The JTC was originally spun off from the EDB with a $100 million grant from the government. (What follows in this and the next two paragraphs is based on interviews with a senior JTC official in the winter of 1995–1996.) It is the physical arm of industrial development in Singapore. The JTC owns about 10% of all land in Singapore and is responsible for infrastructure development in its industrial estates. It finances and builds roads, sewer lines, drainage facilities, and factories with rentable space, which is rented to industries receiving promotional privileges. Once roads, sewer lines, and drainage facilities are constructed, the JTC turns them over to the proper authorities (for roads, sewers, and drainage). Because the JTC is now totally self-financed, tenants are charged the full cost of infrastructure services provision, including the cost of electricity and water.

Before an industry locates in one of the JTC's factories, the JTC carries out an upfront screening of the industry's investment application with the EDB. This includes surveying the potential site of the new factory; making sure that all zoning requirements have been met; and, if investors want to erect their own building, making sure the new building is not an eyesore. In addition, the JTC performs an audit to ensure that water and electrical supplies will be sufficient, to assess whether there will be a need to recycle water, and to ensure that existing sewer lines can handle wastewater discharges. The JTC commits itself to completing these tasks and responding to potential investors within nine days of receipt of their application for space in one of the JTC's industrial estates.

Ultimately, however, access to space in the JTC's factories with rentable space is dependent on access to promotional privileges from the EDB. When the EDB was promoting labor-intensive industry, the JTC granted leases to such factories. As the EDB shifted promotional privileges to activities with higher value added and that were more skill intensive, the JTC limited the allocation of factory space to low-technology, labor-intensive activities (e.g., electroplating and textile dyeing) and began allocating space on the basis of value added per square meter.[17] Now the JTC is trying to maximize the value added per square meter of its factory space. It requires firms to bring in a minimum of $650 of new investment per square meter for factory space and $1,500 per square meter for space in its newer business parks, which have been set aside for headquarters operations and regional operating centers of multinational corporations. Because the JTC works closely with its clients, it also appears to profit from embedded autonomy.

As part of the government's efforts to create an industrial-world oasis among the developing-world economies of Southeast Asia, it has made extensive efforts to make Singapore environmentally responsible (Lee 2000, 174). These efforts included a substantial tree-planting campaign; an early antispitting campaign; and the movement of street hawkers to market areas with piped water, sewers, and garbage disposal. They also included cleaning up the Singapore River and the Kallang Basin, which had become little more than an open sewer fouled with a large volume of household and industrial wastewater. To demonstrate that he was serious about cleaning up Singapore's pollution, the prime minister created the APU in 1970, and the government created the ENV in 1972. Given the importance attached to an environmentally responsible Singapore as part of the government's leapfrogging strategy to industrial-world status, it is important to ask if and how environmental considerations were integrated into this rather elaborate industrial policy machinery.

Answers to these questions were gained by conducting interviews with officials of the EDB, JTC, and ENV in the winter of 1995–1996. The picture that emerges suggests that the government took a series of discrete steps to integrate environmental considerations into the institutions of industrial policy. It invested heavily in building a tough, competent command-and-control regulatory agency. It granted this agency the legal authority, budget, and technical capacity to mount a rigorous inspection and enforcement program. It also granted this agency complete authority over wastewater infrastructure, particularly sewer lines and sewage treatment plants. As will be seen, this created additional opportunities for the ENV to manage industrial pollution. But most important of all, the ENV functioned, from an early date, as an essentially coequal partner with the EDB and JTC. This gave the ENV an important seat at the industrial policy table. Why and how did this happen?

It is clear that this has happened very much because of Prime Minister Lee, who took an early, active interest in the environment after the deterioration of environmental quality in the 1960s (Ferris 1993, 185). This decline was the result, among other things, of smoke from the incineration of plywood waste and water pollution from textile dyeing. As one ENV official stated, if textile firms were dyeing and bleaching blue jeans, the rivers turned blue; if they were dyeing military uniforms, the rivers turned green (interviews, 1995–1996). The prime minister also took an active interest in birds.[18] And following a visit to Boston in 1970, where he noticed that cars were lining up at gas stations to get inspected, the prime minister created the APU within his office (Lee 2000, 181). He put it there because he wanted it to be very close to the EDB so that the environmental impact of new investments attracted by the EDB could be evaluated before the offering of promotional privileges.

Thus, environmental responsibility was an integral part of the prime minister's vision to make Singapore an industrial-world oasis in Asia. Like his counterpart Park Chung-Hee in South Korea, Lee recognized that the PAP's and Singapore's well-being were all tied up with his vision of a prosperous, Western- and market-oriented country. Given this vision and his subsequent commitment to it, the answer to the question of how this was accomplished depended on the actions of bureaucrats in Singapore's strong, autonomous state.

Creating a Command-and-Control Agency

Following the establishment of the APU and ENV, the government set out to create and sustain a tough command-and-control environmental agency. The Clean Air Act (CAA) was passed in 1975; and Clean Air Standards Regulations (CASR)—which specified acceptable levels of air emissions and permitted the placement of those whose emissions were above specified limits on pollution reduction schedules—were announced in the same year (Ferris 1993, 178–179). As Ferris states, it seemed that the CAA granted endless discretion to a director of air pollution control in the ENV to regulate air pollution. Polluting industries were put on a scheduled list, and industries on this list had to obtain written permission from the director before occupying or operating the polluting facility (Foo et al. 1995, 49).

The CAA also gave the director the administrative discretion to refuse permission to occupy or operate a facility. It required facilities to get written permission before they could alter their production process; install, alter, or replace fuel-burning equipment; or alter either fuel type or the height of chimneys (Lin 1997, 258). In addition, the director could require facilities to use certain types of fuel, install and operate certain types of pollution control equipment, install monitoring instruments, change methods of operation, or dismantle the entire facility and cease production (Foo et al. 1995, 50; Ferris 1993, 179). Failure to comply was punishable by a maximum fine of 10,000 Singapore dollars for the first violation and a maximum of 500 Singapore dollars a day for subsequent violations of provisions of the CAA or of conditions imposed on polluters by the director. Those in violation of the CAA or of conditions imposed on them by the director were also subject to jail sentences.

The Water Pollution Control and Drainage Act (WPCDA) of 1975 and Trade Effluents Regulations (TER) of 1976 govern water pollution control efforts. (What follows is taken from Foo et al. 1995.) The WPCDA makes it illegal to discharge certain types of effluent into the country's rivers, lakes, and reservoirs. The penalty for a first offense can include either a fine not exceeding 10,000 Singapore dollars or prison. Subsequent violations are subject to up to six months in prison. Offenders are also subject to strict liability, and no proof of fault is required. The TER specifies a list of allowable discharges into sewers and waterways. The list of covered effluents includes biological oxygen demand, total suspended solids, pH, temperature, and 32 other parameters (Hui 1995, 30–31). Persons or organizations convicted of three offenses against the TER may be required to cease production.

Under the WPCDA, the director of the Pollution Control Department (PCD) of the ENV has the authority to require that polluters meet emissions requirements of the WPCDA and TER before they dispose of wastewater into the public sewer system. This requires operators of new industrial facilities to get permission from the director of the PCD before building near a sewer line. In cases where facilities are constructed before the availability of public sewer lines, the director of the PCD requires the operator of the facility to build temporary treatment facilities (Ferris 1993, 182). Once sewer lines are extended to the new facility, operators of the facility are required to demolish their temporary wastewater treatment facilities (interview with a JTC official, 1995–1996).

The ENV has equally tough regulations governing toxic industrial wastes. The importing, transporting, storage, and use of toxic chemicals is governed by the Poisons Act and the Poisons and Hazardous Substances Act. (What follows is drawn from Ferris 1993.) These acts require that anyone engaged in importing, transporting, storing, using, and disposing of poisons and hazardous substances must obtain a permit from the ENV. In addition, the transport of hazardous chemicals in excess of certain amounts requires the approval of the ENV. The ENV may impose requirements on the packaging, the route and timing of transport, and allowable loads (Hui 1995, 20).

Unlike elsewhere in East Asia, these legal and administrative requirements have been backed up with a substantial monitoring, inspection, and enforcement capability in the ENV. This, no doubt, reflects the pragmatic actions of a very results-oriented government and bureaucracy. The ENV has a telemetric network of 15 air quality monitoring stations that are linked to it by telephone (Hui 1995, 17). Those stations continuously monitor particulate matter, sulfur dioxide, nitrogen dioxide, carbon monoxide, and lead (Hui 1995, 17). Water quality is monitored monthly in the 47 streams and 13 reservoirs that constitute Singapore's catchment area for drinking water. (The ENV also monitors rivers outside catchment areas monthly; Hui 1995, 18.)

Monitoring programs are combined with substantial inspection and enforcement programs. The ENV routinely inspects industrial facilities for compliance with the country's air, water, and toxics regulations. Unannounced or surprise inspections are common. The frequency of inspections is linked to an assessment of a facility's pollution potential. Those with higher pollution potential are inspected more frequently. Although facilities of multinationals are inspected about once a month, those suspected of violations can and have been inspected as frequently as two times a week (Lin 1997, 260). Surprise inspections also cover all permit holders for chemical stores and those transporting hazardous substances. Sometimes, the ENV conducts unannounced road inspections of those who transport hazardous substances, in cooperation with the Registry of Vehicles and Fire Safety Bureau (Hui 1995, 20).

Because of its close work with the JTC, the ENV has been able to develop several additional, unique pollution management opportunities. The ENV's responsibility for sewer lines and sewage treatment plants enabled it to install pH meters in sewer lines at the last manhole cover just outside of each of the JTC's rentable-space factories.[19] If industrial effluents exceed pH emissions standards, the sewer line automatically closes and wastewater discharges back up. When this happens, the ENV is called to reopen the sewer line, and it can sanction pH violaters (interviews with ENV officials, 1995–1996). The ENV has also taken advantage of the fact that the JTC locates similar activities (e.g., food processing, motor workshops, wood processing, and electronics assembly) in similar areas or the same factories. For example, the ENV has installed oil and grease silt traps and centralized oil receptors in factories that house motor repair shops (Lin 1997, 257).

Some limited sense of the scope of the ENV's inspection and enforcement programs can be gained by reviewing its actions in 1994.[20] In that year, it carried out 2,277 assessments of the environmental effects of proposed new factory sitings. It completed detailed impact assessments for a catalyst manufacturing plant, an aromatic plant, and an animal wastes recycling plant (PCD 1994, 5–6, 8, 12, 14). The PCD carried out 34,810 air emissions inspections and approved 181 new control devices, including 60 bag dust filters and 40 scrubbers. It also conducted 566 source tests on emissions, found 8 facilities that failed to meet standards, and required each to take remedial action. In addition, it analyzed more than 1,800 trade effluents for water pollution, and 230 of these failed to meet standards. The ENV required offenders to take remedial action, and it began prosecution of those found in gross violation of standards.

The ENV also approved the installation of 114 water pollution control devices, including 40 oil interceptors, 21 balancing tanks, and 10 sediment tanks. The ENV issued 655 poison licenses, processed 20,194 declarations for importing hazardous chemicals, granted 233 transport approvals of hazardous substances, and carried out 588 surprise inspections of chemical stores. Of these, 82 were found in violation, and the ENV issued 10 verbal and 69 written warnings. It is not surprising that these tough inspections and enforcement programs have resulted in good ambient air and water quality. Between 1985 and 1994, concentrations of sulfur dioxide, particulate matter, and nitrogen dioxide in the air were all below U.S. EPA standards. In addition, all of Singapore's rivers and reservoirs support aquatic life. Dissolved oxygen is above 2 micrograms per liter (μg/l) in 90% of catchment area streams, biological oxygen demand is about 10 μg/l in those streams, and total suspended solids is less than 200 μg/l in those streams (PCD 1994, 16, 17, 23–25, 30).

Yet the management of industrial pollution attending the high-speed growth of manufactures and manufactured exports did not stop with the creation of a tough and competent command-and-control environmental regulatory agency. (What follows, unless noted, is based on interviews with EDB, JTC, and ENV officials in the winter of 1995–1996.) This agency—the ENV—was intimately linked with the premier institutions of industrial policy, particularly the EDB and JTC. As officials of each agency stated, the EDB is responsible for attracting the right kinds of industries. The JTC is responsible for providing factory space and physical infrastructure, and the ENV is charged with ensuring that promoted industries are not too polluting and that they meet the country's tough emissions standards. In the early days, the main role of the APU and ENV was to screen new investments for environmental implications.

At this point, ENV officials said they were very pragmatic. They knew that Singapore had to attract foreign investment in manufactures to grow, and they looked for ways to do this without undermining environmental quality. Because it was not clear what could be done to control pollution, the ENV sent people all over the world to see what was possible. From this, it developed a priority list of control hardware and software suppliers that was given to potential investors. It used this information to set emissions standards for air and water. Initially, standards were set below international best practice levels, but this at least signaled to polluters that they would have to start cleaning up. At the same time, they communicated that standards would be tightened over time. They also gave industrial polluters time to comply and, if asked, they offered help. This suggests that even the ENV may have benefited from embedded autonomy.

Pursuing a Three-Hurdle Strategy

Over time, ENV officials say that they have gotten tougher by asking these kinds of questions: Is the production process proposed by the new investor state of the art? Does it have clean technology built into it? Does the investor plan to practice recycling and waste minimization? As Singapore moved from attracting low-skill, labor-intensive industries to high-skill, high-value-added industries, the EDB and ENV developed a three-hurdle strategy to attract industry. Because Singapore is so dependent on imported and scarce freshwater, the first hurdle was water usage. Both the ENV and EDB looked for industries that saved water or at least recycled it. The EDB even offered promotional privileges, in the form of accelerated depreciation, for purchases of pollution control equipment that recycles water.

Because Singapore is a small island, the second hurdle was to ask what kind of waste the industry created, and whether that waste could be safely disposed of. The ENV is particularly concerned about toxic chemicals that can cause great harm or leach into the country's limited supplies of freshwater. Because of this, the ENV, with the EDB's support, has rejected several applications to locate particularly toxic activities in Singapore. The third hurdle was to ask whether the industry had high value added and was skill intensive.

But how were these environmental considerations ultimately integrated into the industrial promotion activities of the EDB and the infrastructure activities of the JTC? This integration was described by officials of the EDB, JTC, and ENV. (What follows, unless noted, draws on interviews with ENV, EDB, and JTC officials conducted in the winter of 1995–1996.) Once the EDB identifies an investment that is eligible for promotional privileges, it requires the investor to describe the production process that will be used to create a plant's final product. This description must include the amount and kinds of raw materials used, the amount of each that will be embedded in the final product, and the amount of waste and pollution by type. Along with this, the investor must provide a description (including the type of control equipment that will be used) of how waste and emissions will be reduced to meet the country's standards.

This information is sent to the ENV's engineers for evaluation. The primary purpose of this early evaluation of proposed new investments, including expansions of existing plants, is to weed out, at an early stage of development, technologies that will not be able to comply with regulations (Hui 1995, 15). The ENV also uses this planning information to extrapolate mass air and water pollution loadings, and it uses these extrapolations to develop estimates of the impact of a new investment on ambient air and water quality. If the ENV fails to approve the proposed investment or abatement plan and the ENV and EDB disagree, the ultimate decision goes to the cabinet for resolution.

If the ENV approves of the proposed investment and abatement plan, then the EDB begins offering promotional privileges and the JTC begins deciding where to locate the proposed plant.[21] The JTC uses the information provided by investors to the ENV to shrink its hazardous zone by locating plants that use toxic substances in the same area. While this is happening, the JTC examines whether the investor has met all zoning requirements and whether the investor meets the JTC's requirements for value added per square meter of factory space.[22] The JTC also acts as an intermediary for investors by getting approvals from other agencies.

Once the JTC has all of this information, its board (which includes the head of ENV) makes a siting decision (Lin 1997, 254–257, 263). These siting decisions in JTC's industrial parks have to be consistent with the country's master plan for space use.[23] Both the ENV and JTC use the master plan, which has extensive environmental requirements, to decide where and how to locate industrial facilities. Land use in zones is governed by a principle of predominant use (e.g., zones of industrial, commercial, and residential uses). When actual use deviates from predominant use, the government intervenes to force relocation of activities. In the past, this has been used to force relocation of motor repair shops, to move polluting industries along the Singapore River, and to close both a major power plant and a small oil refinery.[24] The density of industrial activity is guided by a maximum plot ratio, defined as the total floor space of a building divided by the area of the property. Ideally, the JTC aims for a maximum plot ratio of two. In addition, the ENV classifies all industrial activity by its impact on the environment. The classification scheme includes clean industries, light industries, and a list of 57 "special" or heavily polluting industries (e.g., oil refining, power generation, and petrochemicals).

The ENV's classifications are used to site industries and fix buffer zone requirements among manufacturing activities, residential and commercial areas, and freshwater sources. The most polluting industries, such as petrochemicals, are placed the farthest from residential and commercial areas; they are never located near a source of freshwater. For the most part, this has meant locating heavily polluting industries on the coast or offshore islands that have been reclaimed by the JTC. Medium and light industries are located between these heavily polluting industries and residential and commercial areas. But the master plan requires buffer zones between medium and light industries that must be respected. Depending on the activity, light industries must be located between 50 and 100 meters from residential or commercial areas. Particularly noisy activities are subject to noise abatement requirements.[25] Others, such as brick making, have to be located at least 1 kilometer from residential and commercial areas (Lin 1997, 255).

If an investor decides to erect his or her own building, the plans must be approved by the Building Control Division (BCD) of the Public Works Department and the Central Building Plan Unit (CBPU) in the ENV. The CBPU inspects building plans to ensure that required environmental facilities are included in building design (Hui 1995, 16). Once construction is completed, the ENV inspects the premises to ensure that environmental facilities have been installed and are properly operating before informing the BCD of clearance by the ENV so that the BCD can issue a temporary occupancy permit and a certificate of statutory completion. After the temporary occupancy permit and the certificate of statutory completion are issued, a facility can begin to operate.

When asked how this works in practice, ENV, EDB, and JTC officials stated that this whole process requires a lot of back-and-forth discussion among their agencies and with clients. But all agreed that this back and forth, which was imposed on them by the prime minister very early in Singapore's industrial development, made it possible to grow rapidly without experiencing long-term deterioration of ambient air and water quality. All also agreed that this would not have been possible if the ENV had not developed into a tough, competent, pragmatic, but fair command-and-control agency and if the EDB, JTC, and ENV had not developed good relationships with their clients in promoted industries.

Conclusions

What are the lessons of Singapore's approach to industrial pollution management for other governments in Asia or elsewhere that are searching for ways to ameliorate the environmental consequences of rapid industrialization? Four lessons seem particularly important. First, Singapore's experience suggests that it is possible to attract significant amounts of foreign investment in manufacturing while imposing tough environmental requirements on those investors. In fact, Singapore started doing this at a time (in the early 1970s) when most of the academic literature on foreign investment focused on pollution-haven explanations of foreign investment.[26]

This foreign investment proved possible because the government of Singapore created an industrial-world oasis in Southeast Asia. Singapore may be the most environmentally responsible city in Asia, if not the world. Its telecommunications and transport facilities, including its airport, roads, ports, and subway system, are as good as any in the world. Its health and education facilities are equally good. Its labor force is relatively docile, hardworking, and better and better educated. In addition, its civil service is pragmatic, goal oriented, responsive, helpful, and notably free of corruption (Lee 2000, 157–173). These characteristics make Singapore an ideal haven for multinationals. To this must be added its unique geographic position on one of the world's most busy shipping lanes.

Interviews with representatives from several multinational corporations in Singapore suggest other factors that explain why Singapore was able to attract multinationals and hold them accountable for their environmental performance. As those officials said, Singapore's ENV made their environmental expectations clear. ENV officials were transparent in their dealings with multinational corporations. They were extremely competent, pragmatic, and fair; yet they were also tough.[27] As one official said, "we knew what was expected and we trusted them. We may not have liked it, but given the other advantages offered by Singapore, this was just one of the costs of doing business."

The second lesson of Singapore's approach is that it was predicated on building and sustaining a pragmatic, tough, competent, and fair command-and-control environmental agency, the ENV. Initially, emissions standards set by the ENV were below U.S. EPA and European standards. This was done to signal to polluters that they would have to meet some standards even if they were not very tough. The ENV also made it clear that over time the ENV would adopt U.S. EPA and European emissions standards. Pragmatic and tougher and tougher emissions standards were complemented by strong programs to monitor ambient air and water quality. To this, the ENV added a rigorous, honest, and fair inspection and enforcement program. This pragmatic step-by-step process should be particularly attractive to new environmental agencies trying to establish their credibility.

The third lesson is that the government used its control over land use and industrial estates to good advantage. Access to a factory with rentable space in an industrial estate was limited to investors that could meet the ENV's emissions standards. Housing similar activities in the same estate made it possible to develop several unique strategies to control industrial pollution. Siting the most polluting activities in areas farthest from where people lived also helped to minimize the potentially harmful environmental effects of manufacturing activities. Because many developing countries in East Asia and elsewhere have industrial estate authorities that offer space and good infrastructure facilities, it should not be too difficult to build at least some environmental considerations into industrial estate management. At the very least, this provides a focus for monitoring and enforcement activities.

The fourth lesson is that the government gave its strong command-and-control environmental agency an important seat at the industrial policy table. The creation of the APU in the prime minister's office in 1970 signaled to the EDB and JTC that he was committed to economic growth through attracting foreign investment while protecting the environment. Because of this, the ENV came to play an important role in screening potential new investment for its environmental implications. It occasionally rejected investments that were too polluting. It required investors to plan for abating pollution by requiring that abatement plans be part of an investor's investment package. These plans were required before the EDB could offer promotional privileges and before the JTC could provide factory space. And the ENV required investors to install abatement equipment that would meet the country's emissions standards before granting an industrial facility a temporary occupancy permit or a certificate of statutory completion that would enable it to begin operating. Because most developing countries tend to have some kind of investment promotion agency, they might consider linking promotional privileges to at least limited environmental requirements.

It is equally important to note what is unique about Singapore that most likely limits the wholesale transferability of its pollution management policies to others in East Asia. And it is important to point out the more and more obvious limitations of Singapore's excessive reliance on technology-based command-and-control policies. To begin with the latter, there is growing evidence that Singapore's command-and-control approach to pollution is becoming too costly, and there is some evidence that the ENV is considering less costly alternatives (Ling 1994).[28] With respect to the former, few countries in the world have such strong, autonomous, pragmatic, and goal-directed governments. In addition, in most places where strong, autonomous governments have gone unchallenged by either the private sector or other actors in civil society, rent-seeking policies rather than developmentally oriented ones have resulted.

This raises an interesting question: Why has Singapore's strong, autonomous government been so committed to what might be labeled sustainable development rather than rent-seeking? As was hinted at above, much of this reflects the vision of the prime minister and his party. Together, they concluded that the best prospects for their own well-being, and for the well-being of Singaporeans and the country, lay in creating an environmentally responsible Singapore economically linked to the West. This, no doubt, reflects the dire circumstances of Singapore in the 1960s: Federation with Malaysia had collapsed, the British military was withdrawing, confrontation with Indonesia was well under way, and the PAP faced a strong challenge from the left. Faced with these events, and several years of disheartening trial-and-error alternative approaches to economic development (Lee 2000, 57), it is not surprising that bourgeois elements within the PAP would opt for such a strategy. This strategy fit their basic worldview. As Prime Minister Lee has stated, he and the PAP categorically rejected prevailing dependency arguments about the costs of linking with multinationals and exporting to the West. This, no doubt, imparts a unique character to Singapore's experience.

In addition, few countries in the world have been as successful as Singapore at land-use planning. Singapore's success in this area is undoubtedly due to the fact that it is a small city-state that has made the most of its embedded autonomy. Most governments, even of cities, lack these characteristics. Similarly, because of its small size, Singapore has been able to build sewers for all industrial activities and virtually all households. This is probably beyond the reach of larger, more populous countries. Also, the city-state of Singapore, unlike its neighbors (particularly Indonesia, Malaysia, and Thailand), has not had to contend with the difficult pollution problems associated with large- and small-scale agroprocessing activities. At the very least, this meant that Singapore's ENV did not have to deal with rising water pollution loads from crude palm oil processing mills, as in Malaysia, or with air pollution loads from forestry activities, as in Indonesia.

Finally, because Singapore relied on multinational corporations for its growth, it did not have to contend with growing a domestic business class, as did Indonesia, Malaysia, South Korea, Taiwan, and Thailand. This both freed the government from business interests and linked it to environmental developments in the OECD countries. Thus, over time, the multinational corporations that located in Singapore and dominated its economy came to view their environmental relationships with the Singaporean government much as they did their relationships with their home governments. Because these governments all had tough environmental requirements, it was easier for the government of Singapore to manage its environmental relationships with these corporations.

Notes

[1]Singapore's drive to attract manufacturing multinationals was led by the Economic Development Board, an investment promotion agency created in 1961. Between 1961 and 1965, the board promoted import-substitution industries (Low 1993, 63). Singapore's initial export orientation and economic takeoff occurred between 1966 and 1973 (Hughes 1993).

[2]In that year, total suspended particulates averaged 32 micrograms per cubic meter ($\mu g/m^3$) of air; smoke levels in ambient air, which averaged 24 $\mu g/m^3$, were well below U.S. Environmental Protection Agency (U.S. EPA) standards (PCD 1980, 8–9).

[3]Concentrations of smoke in the air averaged 23 $\mu g/m^3$; total suspended particulates averaged 34 $\mu g/m^3$ (Tay 1993, Table 4). Both are well below the U.S. EPA standard, as are concentrations in ambient air of suspended particulates (48 $\mu g/m^3$), sulfur dioxide (19 $\mu g/m^3$), and nitrogen dioxide (29 $\mu g/m^3$) (PCD 1994, 24). Ambient raw water quality, although not quite as good as in the United States, is also improving. Dissolved oxygen is above 2 micrograms per liter of water (μg/l) 91% of the time, and biological oxygen demand (BOD) on raw water is a little less than 10 μg/l—about 68% of the time (PCD 1994, 30).

[4]For a discussion of widely publicized pollution incidents in the development of environmental improvement strategies in Taiwan and South Korea, see Lee and So (1999), Tang and Tang (1999), and Eder (1996).

[5]The government has routinely shut down distribution of the international press in Singapore when it publishes stories unflattering to Singapore (Means 1998, 103–105), and it has gone to great lengths to defend Asian values (Lee 2000, 490–496). In addition, because Singapore launched its environmental programs so early, its government has not been subject to international pressure to clean up the environment.

[6]Exports of manufactures increased from $56 million in 1968 to $49.5 billion in 1992 (World Bank 1990, 490–491; 1994c, 578–579).

[7]For a discussion of Singapore's institutions of industrial policy, see Times Academic Press (1993), Huff (1994, 1995), Findlay and Wellisz (1993), Krause (1988), and Lim (1983).

[8]The country's geographic location astride a major shipping route between the Indian and Pacific oceans enabled Singapore to become, by 1950, a major international seaport and a center for telecommunications, air transport, and mail for the Far East. At this time, Singapore was also the home of the world's largest market for natural rubber, the largest international futures market for tin, and a major oil distribution center (Huff 1994, 31). And it was the only place in Asia with a substantial middle class. Per capita income in 1956 was a third that of the United Kingdom, the infant mortality rate was low (41.4 per thousand live births), and car ownership was fairly widespread (30 people per car, versus 120 for the rest of Asia) (Huff 1994, 33).

[9]From 1966 to 1980, the PAP won all the seats in parliament. Opposition political parties captured one seat in 1981, two seats in 1984, one seat in 1988, and four seats in 1991 (Means 1998, 101).

[10]The opposition party winner of a seat in parliament in 1981 was subsequently expelled from parliament for public criticism of the government (Means 1998, 101). Another opposition party member was detained for 26 years, without trial, under Singapore's Internal Security Act, and a third who ran for election in the 1988 elections was arrested under the Internal Security Act and subsequently barred from parliament for alleged violations of the country's tax laws (Means 1998, 102).

[11]In 1965, wage costs were thought to be 20% to 30% too high for world markets (Huff 1995, 1424).

[12]By 1969, hourly labor costs were one-eleventh of U.S. levels and lower than in Hong Kong, South Korea, and Taiwan. At the same time, labor productivity in Singapore equaled that of the other East Asian newly industrializing economies (Huff 1995, 1424).

[13]Lim (1995, 304–312) reviews the print media's coverage of the environment in Singapore. He argues that coverage does little more than reinforce government policies and report on positive events (e.g., Earth Day). He says the press is very careful not to criticize the government or government officials.

[14]For a discussion of how this works for the few environmental NGOs extant in Singapore, see Lim (1995).

[15] For a discussion of the role of pressure from neighbors, communities, and environmental NGOs in environmental improvement programs in East Asia, see Lee and So (1999).

[16]Silcock (1985, 293) has suggested that no more than 50 senior civil servants are responsible for Singapore's development success.

[17]In fact, it is highly unlikely that the JTC will grant any additional factory space to these kinds of activities (interview with JTC official, winter 1995–1996).

[18]At one point, he is reputed to have asked one of his senior government officials where all the birds had gone. The official replied that the birds were disappearing because many trees had died or been cut down. The prime minister then ordered a substantial tree-planting campaign (personal communication with Robert Wade, November 3, 2000).

[19]It should also be noted that the ENV is committed to providing sewer lines to all industries, commercial establishments, and households. By 1991, the ENV had invested 1.7 billion Singapore dollars in more than 2,000 kilometers of sewer lines, more than 100 pumping stations, and 6 sewage treatment plants (Khoo 1991, 129).

[20]Ideally, these actions should be placed in a relative context. For example, one would like to know whether 566 source emissions tests were large or small relative to economic activity. Unfortunately, data limitations made it impossible to place these numbers in a relative context.

[21]In the early days, the ENV gave the JTC "hell," particularly for not expanding sewer lines fast enough to keep pace with industrial estate and new plant expansion. When this happened, the ENV required the JTC to build temporary treatment plants for industrial wastewater; and then, when sewer lines caught up, the ENV required the JTC to demolish the temporary treatment facilities and to hook up to the sewer lines (interview with JTC official, winter 1995–1996.)

[22]As part of Singapore's efforts to attract more skill-intensive and higher-value-added activities, the JTC is trying to maximize the value added per cubic meter of factory space in its industrial estates (interview with JTC official, winter 1995–1996).

[23]This master plan has zoned the entire island for particular land uses. In this way, environmental considerations are also integrated into land-use planning.

[24]Because of this, about 50% of the JTC's tenants, by number of tenants, are small workshops that have 1 to 15 employees and are particularly polluting. This includes motor repair shops that use a lot of grease and oil and others that use cleaning solvents (interview with JTC official, winter 1995–1996).

[25]Continuously noisy activities must reduce noise levels to 75 decibels between 7:00 a.m. and 11:00 p.m. and 65 decibels between 11:00 p.m. and 7:00 a.m. (Lin 1997, 257).

[26]The pollution-haven hypothesis argues that foreign investors in rich countries respond to tough environmental regulations in rich countries by moving their dirtiest industries out of countries with tough environmental regulations to those countries (particularly developing countries) with weak or nonexistent environmental agencies. For a discussion of the pollution-haven hypothesis, see Leonard (1988).

[27]As one representative from a U.S. multinational stated, "some companies have been barred from bidding on government contracts because of their poor environmental behavior." This official also stated that even if he has not seen one of the ENV inspectors for a while, he knows they will come and that they are incorruptible (interview, winter 1995–1996).

[28]In Chapter 7, I suggest what steps Singapore might take to achieve more cost-effective industrial pollution control.

3

Democratization, Industrial Policy, and Pollution Management in Taiwan

Pollution management policies in Taiwan have been shaped by the government's responses to internal and external shocks—particularly democratization and the loss of international recognition.[1] Democratization, which commenced in 1986, forced the ruling party, the Kuomintang (KMT), and the strong autonomous state it controlled to respond to public pressure to clean up the environment (Tang and Tang 1997, 1999). The loss of international recognition made the government of Taiwan sensitive to international criticism of its domestic policies, including its environmental policies (interviews with officials of the Council for Economic Planning and Development in Taiwan, November 1995). Pressure from overseas Chinese, particularly in the United States, to clean up the environment added to the growing calls for the government to stop the environmental deterioration that attended high-speed industrial growth. All of these contributed to an effective governmental response to environmental deterioration that resulted in significant improvement in ambient environmental quality, particularly air quality (TEPA 2000). Because the KMT and the public-sector institutions it created to manage Taiwan's equitable growth loom so large in the evolution of environmental management, the story of industrial pollution management begins with the KMT and its economic and industrial policies.

Historical Developments

Neither economic policy nor industrial policy in Taiwan can be neatly separated from Taiwan's and the KMT's history. Economic and political crises (in China and Taiwan), ideology (particularly the philosophy of Sun Yat-Sen), and state–society relations (between the KMT and the indigenous Taiwanese) all exerted influence on development (and industrial) policy-making.[2] Once forced to retreat to Taiwan, the KMT vowed not to permit the same forces that had contributed to its defeat on the mainland to reappear. Chief among these were landlordism, belligerent trade unions, independent bankers who fueled inflation, a government beholden to vested interests, and weak party discipline (Wade 1990, 260).

The Kuomintang's Role

Over time and in turn, the leaders of the Kuomintang responded to each of these threats, but how they did so was affected by ideology and state–society relations. Ideologically, the KMT was wedded to the economic philosophy of Sun Yat-Sen. Sun's three principles and KMT ideology were an odd mixture of private ownership of capital, central (indicative) planning, and socialism (concern for people's livelihood and equity in income and wealth distribution) (Haggard 1990, 77). These principles are enshrined in the constitution of Taiwan, which states:

> National economy shall be based on the principle of People's Livelihood and shall seek to effect equalization of land ownership and restriction of private capital in order to attain a well balanced sufficiency in national wealth and people's livelihood ... With respect to private wealth and privately operated enterprises, the State shall restrict them by law if they are deemed detrimental to a balanced development of national wealth and people's livelihood ... Private citizens' productive enterprises shall receive encouragement, guidance and protection from the State. (Wade 1990, 260)

Unwillingness to rely on unbridled market development can also be seen in one of the economy's five-year plans, which states:

> The main characteristic of private enterprise—the profit incentive—will be preserved and the weakness of private enterprise—concentration of wealth—can be avoided ... For private enterprise ... (the state will protect) ... reasonable personal income ... and freedom of economic enterprise ... However, ... manipulation of society's economic

> lifeline in the hands of a few and over concentration of wealth will not be allowed. Consequently, the government must take part in all economic activities and such participation cannot be opposed on the ground of any free economic theory. (Wade 1990, 261)

Thus the KMT, like its counterparts in Japan, Singapore, and South Korea, recognized a preeminent role for the state in guiding market development to meet avowedly political ends. But precisely how was the state to guide market development in agriculture, industry, and foreign trade?

Part of the answer can be found in the way state–society relations developed. As is well known, the KMT was an alien political force in Taiwan with little organic connection to Taiwanese society. Because of this, it was relatively easy for the KMT to structure state–society relationships to enhance the state's autonomy from organized groups in civil society. It did this in a variety of ways. State-controlled youth groups and labor unions preempted the development of independent student organizations and trade unions (Haggard 1990, 81). In agriculture, a thoroughgoing land reform destroyed the landlord class at the same time that the KMT turned a Japanese-inspired network of rural associations to its political advantage. In industry, the state relied on state-owned enterprises in the commanding heights of the economy, virtually state-owned banks that controlled access to cheap commercial credit, and deliberately small-scale industries that were beholden to the centralized state to limit political power of private industry.[3] When combined with a Leninist-like party that reformed and purged itself of undisciplined elements in the early 1950s, the net result was, as in Singapore, a strong, autonomous state and weak organizations in civil society (Wade 1990, 235, 242).

How were Sun Yat-Sen's economic philosophy, state autonomy, and a reformed party used to promote well-balanced sufficiency in national wealth and people's livelihood in agriculture? Following a land-reform program that culminated in returning land to the tillers in 1953, the government reorganized agricultural input and output markets, including credit and fertilizer markets, and invested heavily in rural infrastructure and rural industrialization.[4] Government control of agricultural markets facilitated heavy taxation of agriculture. Bartering rice for fertilizer at prices unfavorable to farmers, compulsory sales of rice at low prices, land taxes, and a variety of other devices proved to be effective ways of extracting a surplus from agriculture (Johnston and Kilby 1975, 256–257). Investment in rural infrastructure, which contributed to increases in agricultural productivity, and rapid growth of nonfarm rural employment made it possible to get this surplus without impoverishing the peasantry (Amsden 1979, 353–354). The surplus was used by the government to develop industry—first to serve the domestic market and subsequently to serve export markets.

State Policies and Agencies

What, however, precisely did the state do to promote industry, and how did it do it? Turning first to the second question, industrial policy was and is limited to a small number of agencies and individuals. (What follows draws heavily from Wade 1990, 196–217.) At the top of the system are the president and an informal inner group of the cabinet known as the Economic and Financial Special Group (EFSG). The EFSG consists of the minister of economic affairs, the governor of the central bank, the minister of finance, the director-general of the budget, and several ministers without portfolio. This group is advised by the Council for Economic Planning and Development (CEPD), the Industrial Development Bureau (IDB) of the Ministry of Economic Affairs, and the Council for Agricultural Planning and Development.

Industrial policymaking in Taiwan is dominated by the CEPD and IDB. The CEPD formulates short- and medium-term development plans; but because it has no responsibility for implementation, one might say that the IDB has extensive primary authority over industrial policy. It takes the lead in trade policy and policy reform. It is responsible for implementing a large array of fiscal incentives that accompany the 1960 (and amended) Statute for Encouragement of Investment. It plays a leading role in public-sector policy for research and development, including strong ties to the economy's premier science and technology agency, the Industrial Technology Research Institute (ITRI). It also plays some role in the allocation of subsidized credit from state-owned commercial banks to particular industries and firms. And it relies heavily on state-owned enterprises in the commanding heights of the economy—petrochemicals, steel and basic metals, and shipbuilding. This large sway in industrial policy was summed up by Wade (1990, 202):

> The IDB has authority to draw up lists of items to be given fiscal incentives and the lists of tariffs and import controls; to decide case by case requests for importing items on the "approval" list, and more generally to encourage firms to make purchasing agreements with domestic suppliers; to organize the calculation of input-output coefficients for the duty draw-back scheme; to help to establish orderly export marketing arrangements in industries where cut-throat competition is resulting in buyers complaints; to oversee price negotiations in sensitive sectors like petrochemicals; to grade the production facilities of firms in key industries; to approve applications for loans from various special loan schemes and for loan guarantees; to provide administrative guidance to firms . . . and to be the spear throwers and shock troops and the main point of contact between foreign companies and the bureaucracy.

But being first among equals means that others are involved in industrial policy, particularly that related to exports of manufactures. The Bureau of Commodity Inspection and Quarantine and several commodity-specific testing agencies, such as the Electrical Testing Center, inspect export items (Wade 1990, 144). The China External Trade Development Council, an export-marketing and -promotion agency, boosts Taiwan's exports to foreign markets, organizes trade fairs, and carries out market research for exporters (Keesing 1988). Finally, the government, including the premier, is engaged in an annual program of export awards that acknowledges individual firms (and entrepreneurs) for their export performance.

Because Taiwan's industrial policy is so pragmatic and results oriented, this rather vast array of institutions and incentives has profoundly shaped industrial development. Studies of trends in the income elasticity of demand, technical change, and the current composition of imports continue to help identify industries that might be developed next (Wade 1990, 188). In the 1950s, this meant promoting agroprocessing and light consumer goods. When the economic gains from first-stage import substitution slowed in the late 1950s and early 1960s, industrial policy emphasized exports of labor-intensive manufactures. As industrial development deepened and demand for intermediate goods and capital goods increased during the 1970s, emphasis shifted to heavy chemical industries—steel and basic metals, petrochemicals, and shipbuilding. Finally, as the economy lost its comparative advantage in low-wage and low-skill industries and concern for the environment grew, industrial policy shifted to skill-intensive technological industries and less- and nonpolluting industries.

In each change in industrial policy, Taiwan was a follower of what had occurred earlier in Japan and the West. Its progression up a product ladder that had been pioneered by others provided its blueprint for industrial policy. A slowing of growth among promoted import-substitution industries and increased imports in new areas signaled the need to shift promotional activities from import-substitution industries to export-oriented ones. Later, increased demand for intermediate and capital goods signaled the need to deepen industrial structure. Finally, rising wage rates and tougher environmental standards signaled the need to promote skill-intensive, low-polluting industries.

These developments resulted in amendments to the Statute for Encouragement of Investment, which spells out the criteria for eligibility (by item and industry) for fiscal incentives.[5] It also led to changes in the list of industries eligible for administratively allocated and subsidized credit. When government officials feared that the private sector might be too slow to respond to these fiscal and financial incentives, they turned to state-owned enterprises as in plastics and steel, or to the quasi-public ITRI for the development of a high-tech computer chip industry (Wade 1990, 99–100, 103–108).

As is now known, this approach to economic and industrial development has been profoundly successful. Taiwan's success appears to have hinged, as did that of Japan, Singapore, and South Korea, on developmentally oriented political elites who saw economic and industrial development as serving their own as well as the nation's interest. Consequently, this tiny elite gave large sway to a small group of highly educated technocrats, who adopted a nonideological, trial-and-error, results-oriented approach to economic and industrial development. These technocrats enjoyed, for the most part, a large measure of insulation from the political pressure of vested interest groups in civil society. They also enjoyed, like their counterparts in Singapore, institutionalized channels of communication with the private sector, or what Evans (1995) calls embedded autonomy. This particular political economy of policymaking more or less characterizes the rest of the East Asian newly industrializing economies (NIEs) (Johnson 1987, 136–164).

Industrial Policy Institutions and Environmental Protection

Although there is general agreement that Taiwan's particular political economy did make a difference in "economic growth with equity," less is known about its influence on the environment. Some have suggested that it contributed to substantial deterioration of the environment (Chan 1993, 35–56). Others have argued that it created an environmental disaster (Bello and Rosenfeld 1990, 195–214). Still others have argued that the relative strength and political insulation of the economic ministries contributed to a delayed response to environmental degradation (Chun-Chieh 1994, 23–47). Unfortunately, there has been virtually no rigorous research on these important topics. Even more surprising, however, little or nothing is known about whether this particular political economy has been positively engaged in environmental protection.

There are several reasons to suspect that Taiwan's political economy might make a positive difference in the environment. For one, the first East Asian NIE—Japan—has already done so. At least some comparative environmental data suggest that Japan has outperformed its counterparts in the Organisation for Economic Co-operation and Development (OECD).[6] But this is not the only reason to suspect that Taiwan might also do so. Its systematic and quick movement up the product ladder in industry shows that Taiwan has been particularly adept at capitalizing on the real advantages of being a late industrializer. Perhaps it could do the same regarding the environment.

Late industrialization and openness to foreign capital and technology create several opportunities to leapfrog to less costly, more effective urban industrial environmental outcomes. (What follows is drawn from O'Connor

1994, 32–35.) Because Taiwan is dependent on the OECD economies for inputs of capital and technology, there is some possibility that at least some of these inputs will result in lower pollution and material usage than the current capital stock. Because industrial growth rates in Taiwan are significantly higher than the developing-economy average, this creates more opportunities to access these newer inputs, which produce less pollution. Taken together, this combination—higher industrial growth rates and dependence on the OECD countries for these inputs, along with a demonstrated ability to take advantage of being a later industrializer—could facilitate rapid environmental catch-up in Taiwan.

These, however, are not the only factors that might accelerate Taiwan's environmental catch-up. Because its loss of international recognition makes Taiwan more sensitive to international criticism than most of the other East Asian NIEs, international pressure to clean up the environment is likely to have a greater effect in Taiwan than elsewhere (interviews with a CEPD official in Taiwan, November 1995). The recent and stable transition to democracy has added to international pressure to reduce pollution. The freeing of restraints on the creation of autonomous groups in civil society (Yun-han 1998, 138) exposed the ruling party to a vociferous environmental protest movement and to attacks by opposition political parties hoping to take advantage of the KMT's environmental record and its closeness to polluting industries (Tang and Tang 1997, 1999). This suggests that a vigorous environmental protection program might be an important way for the KMT to contend with opposition political parties for electoral support.

Is there any evidence that the Taiwanese government has responded to domestic and international pressure to clean up the environment? Before democratization and its loss of international recognition, Taiwan, like nearly all the other East Asian NIEs, followed a "grow first, clean up later" industrial development strategy that was export led (Steering Committee 1989). The government's first environmental laws date from the mid-1970s. Although those laws covered specific media—air, water, and solid waste—they had little effect. They were not part of a comprehensive environmental law; no clear emissions or effluent standards accompanied them; and jurisdictional authority for standards setting, monitoring, and enforcement was unclear (Tsong-Juh 1994, 438–441). In 1982, some of the jurisdictional problems were eased with the creation of the Environmental Protection Administration in the Department of Public Health.

The Taiwan Environmental Protection Administration

Subsequently, international pressure from overseas Chinese Americans (interviews in Taiwan, November 1995) and domestic public pressure after democratization led the government to be more forceful (Tang and Tang

1997, 1999).[7] Between 1980 and 1992, the media-specific acts of 1974–1975 were amended and strengthened. The Waste Disposal Act was amended in 1980, 1985, and again in 1988. The Water Pollution Control Act was amended in 1983 and 1991.[8] The Air Pollution Control Act was amended in 1982 and 1992.[9] Following the first steps toward democratization in 1986 and 1987, the government created a cabinet-level Taiwan Environmental Protection Administration (TEPA) and modeled it after the U.S. Environmental Protection Agency (U.S. EPA).[10] By 1993, Article 18 of the constitution stated, "Environmental and ecological protection should be given the same priority as economic and technological development" (Office of Science and Technology Advisors 1995, 1).

Despite the continuing lack of a basic environmental law, TEPA was given responsibility for standards setting, environmental monitoring, and enforcement. In a few short years, TEPA evolved into a respectable environmental agency. Ambient air and water quality standards generally follow U.S. standards.[11] TEPA developed a rigorous process to set emissions standards and an equally rigorous monitoring and enforcement program.[12]

Standards setting depends heavily on expert committees that exclude representation from industry and the IDB.[13] The IDB describes this process as closed to them. As a result, firms often complain to the IDB that the standards are too tough and, in at least one instance, an entrepreneur went directly to the president to complain about a particular standard.[14] In response, the IDB has begun to play a mediating role between TEPA and industry. It now regularly interacts with TEPA, and it is considering developing its own technical capability to assess TEPA's standards. Both senior government officials and prominent academics advising TEPA openly applaud TEPA's tough emissions and effluent standards and see them as needed and effective "sticks" to force compliance.[15]

Monitoring and enforcement appear to be equally tough. (This and the next two paragraphs, unless noted, are taken from TEPA 1993, 60–67.) By 1991, when Taiwan held its first open elections for the National Assembly and the Legislative Yuan, TEPA had 184 ambient air quality monitoring stations to measure particulate matter, sulfur dioxide, nitrogen dioxide, carbon monoxide, and ozone. The results from these stations are aggregated into individual indices that are aggregated into an overall pollution standards index. TEPA now routinely reports the percentage of days in a year or month when this index is below 100 (considered healthy), between 101 and 199 (unhealthy), and above 200 (very unhealthy). It also closely monitors the reliability of the monitoring stations.

In 1989, two years after the lifting of martial law and two years after TEPA was formed, TEPA committed itself to meeting the economy's ambient air quality standards by 2002. To facilitate movement in this direction, it screened more than 9,000 factories, and in those factories it identified 24,343 different air pollution sources, 10,484 pieces of pollution control equipment, and 74,343 smokestacks. Of these screened factories, 6,959 were identified as major polluters (O'Connor 1994, 97). TEPA required each of these factories to draw up and submit environmental improvement plans. Following this, TEPA carried out random inspections of 1,543 smokestacks and ascertained that 23% failed to meet emissions standards. These random inspections were subsequently complemented by two targeted projects: the Flying Eagle Project and Rambo Project. The Flying Eagle Project used police helicopters to respond to citizen complaints about factory emissions, whereas the Rambo Project was a "get tough" effort that randomly rechecked state-owned and private factories that had failed earlier inspections.

These actions resulted in fines, suspensions of operating permits (government-mandated factory closings), and voluntary factory closings. TEPA carried out intensive inspections of the factories that failed to respond to its request for environmental improvement plans. These inspections led to fines for some and to suspension of operating permits for others. In some instances, inspected plants (e.g., the Kaoshiung factory of the Taiwan VCM Corporation) that failed to improve their environmental performance within a specified time period were fined on a daily basis. This led the company to voluntarily suspend operations until it was able to reduce its emissions. In other instances, repeat violaters of emissions standards (e.g., the Hualien plant of the Chung Hwa Pulp Corporation) voluntarily suspended operations until they could meet standards.

In still other instances, factories were closed,[16] or the government banned certain heavily polluting activities or restricted them from locating in densely populated areas. (The examples that follow are from IDB n.d., 4–10.) Alkali and chlorine factories were prohibited from using a mercury electrolysis process, and by 1993 all such processes were replaced by an ion electrolytic membrane process. New polychlorinated biphenyl–based pesticide factories were prohibited, and all preexisting factories began to be eliminated. The production of certain carcinogenic organic dyes were banned. New paper pulp factories were banned from central areas, and they were required to install black liquor recycling equipment.

The aggregate evidence on the use of negative sanctions reinforces these case examples. (What follows is based on data from the Office of Science and Technology Advisors 1995, 27–29.) The number of inspections grew dramatically from about 338,000 in 1988 to 729,000 in 1994. Even though only about 6% of the inspections were for stationary source water and air pollution, this amounted to 40,000 site-specific inspections annually between 1990 and 1994. During this same period, fines for violations of standards increased from 2.66 billion real New Taiwan dollars (NT$) to NT$6.32 billion, while average fines increased from NT$7,869 in 1988 to NT$8,679 in 1994. This represented a 10% increase in the average fine.

TEPA has gone one step further. It has developed an impressive database on the output of various pollutants by industry for 11 industry groups and 3 specific state-owned enterprises—Taipower, China Petroleum Company, and China Steel. These estimates have been used to extrapolate output of pollutants under a no-enforcement, or natural growth, scenario. One measure of the effectiveness of enforcement involves comparison of this natural growth scenario to output of pollutants after enforcement. For example, TEPA estimated the natural growth of total suspended particulates in 1991 at 1.14 million metric tons. Actual emissions after pollution control were estimated to be only 1.00 million metric tons, or a reduction of 11.4% (TEPA 1993, 51). TEPA has developed a similar database for solid and hazardous industrial waste. The sources of such waste are linked to 20 specific industries. They are further subdivided into generalized, toxic, corrosive, and infectious waste.

Other Policies and Programs

Environmental protection, however, neither starts nor stops with TEPA. The government of Taiwan has begun to integrate environmental considerations into industrial policy. This is being done in three principal ways. The government is following an import substitution industrial development strategy for the creation of an indigenous environmental goods and services industry. It is heavily subsidizing industry purchases of pollution control and abatement equipment. And it is financing research on pollution prevention and providing industry with subsidized technical assistance in prevention.

All of this is part of the latest industrial development strategy and the IDB's efforts to finally get engaged with helping Taiwan achieve a less pollution-intensive industrial development path (IDB 1995, 11, 20–21). Appreciation of the exchange rate, rising wage rates, emerging labor shortages, and increased demands for a cleaner environment contributed to an export of industry that some in the IDB feared was leading to a "hollowing out of industry." This, no doubt, contributed to reluctance in the IDB to take the environment more seriously until it was forced to do so by the government. To prevent the hollowing out of industry, the government promulgated a six-year national development plan to upgrade industry and replaced the 1960 Statute for Encouragement of Investment with a Statute for Upgrading Industries. This statute provides selective incentives to firms to purchase automated production equipment and technology, increase expenditures on research and development, improve product quality, increase productivity, reduce energy use, promote waste reclamation, and purchase pollution control and abatement equipment.

The government also began to promote 24 key high-technology, high-value-added items in its Ten Emerging Industries program. The promoted industries included communications, semiconductors, precision machinery, aerospace, and, most notably, environmental goods and services. These industries were selected because they cause little pollution, have strong market potential, are technologically demanding, are not energy intensive, and have high value added (IDB 1995, 28).

The government is relying on several promotional privileges to facilitate the growth of a domestic environmental goods and services industry. Firms in this nascent industry have, by law, been organized into industry-specific associations.[17] Government environmental contracts (e.g., to build public-sector waste incinerators or provide pollution prevention technical assistance to private-sector firms) are reserved for firms in these industry-specific associations. Because the government has adopted an explicit private-sector approach to environmental cleanup, these benefits are likely to be substantial.[18]

Domestic providers of environmental goods and services are also favored by tax, commercial bank lending, and land-use policies. Firms in all of the Ten Emerging Industries, including the environmental goods and services industry, are eligible for either a 20% investment tax credit or a five-year tax exemption, plus a double retaining of surplus earnings (IDB 1995, 29). They are eligible for loans from commercial banks and the Executive Yuan's Development Fund at preferential rates, and they are given priority consideration in the acquisition of industrial land.[19] Local environmental hardware providers benefit from a 20% tax credit that accrues to firms purchasing pollution control and abatement equipment.[20] Firms in this industry are also eligible for export assistance, but little is known about how this program works. If it follows practice elsewhere in East Asia, access to assistance may be conditioned on export performance (Rhee et al. 1984, 22–37). Because the government has established explicit export targets for the environmental goods and services industry through 2002, something like this may be happening.[21]

In addition to this broad array of support for the development of a domestic environmental goods and services industry, the government offers a wide range of programs to assist firms trying to reduce industrial pollution. Some of these programs offer fiscal and financial incentives. Others provide technical assistance, particularly for pollution prevention and waste treatment. The purchase of pollution control and abatement equipment entitles purchasers to tax credits (of either 20% or 10%), and between 5% and 20% of the costs of expenditures on energy conservation and on recycling equipment or technologies can be credited against profits (IDB 1995, 21). A joint IDB–TEPA Waste Reduction Task Force provides free technical assistance to firms on waste reduction or minimization.[22] The IDB also runs an Information Service for Exchange of Industrial Wastes, sponsors demonstration projects, and has a program to enable small and medium-sized enterprises (SMEs) to congregate in industrial parks. The SMEs that move to these parks are provided with good infrastructure, including common wastewater treatment facilities. In return, the SMEs are required to elect an SME committee to enforce emissions and effluent standards within the park (interviews with an IDB official in Taiwan, November 1995).

Finally, the IDB finances a growing research program on clean technologies. It has created a clean technologies unit and contracted its research on clean technologies out to the United Chemical Laboratory (UCL) laboratory of ITRI.[23] Most remarkably, ITRI's clean technology researchers are going well beyond plant-by-plant pollution prevention. They are watching closely what others (e.g., the 3M Corporation and U.S. EPA Toxic Release Inventory) are doing, and they are exploring several cost-effective alternatives to develop policy-relevant estimates of the pollution intensity of output by industry subsector. One of these measures compares the weight of materials used to produce a product with the weight of the final product. Another looks at waste (the difference between the weight of inputs and the weight of final product) per NT$ of sales. A third disaggregates waste into four categories (raw materials, industrial water, energy, and toxic chemicals) per NT$ of value added. A fourth benchmarks Taiwanese industry, such as the water intensity of wafer fabrication, against international best practice.

Although outsiders are likely to view calculating the pollution intensity of highly disaggregated industry subsectors as either prohibitively expensive or too difficult, in Taiwan it appears to be little more than an extension of what the IDB already does to administer its duty-drawback system for exporters.[24] (What follows is based on interviews conducted at the new National Center on Cleaner Production in the Environmental Sciences and Technology Division of UCL in November 1995.) The technical staff of the new National Center on Cleaner Production is doing this for two reasons. First and most important, they hope that better understanding of the pollution intensity of production processes will enable them to redesign those processes to reduce their pollution intensity. Because they see this as too risky for the private sector, scientists at ITRI view this as an important role for government.

Second, the technical staff see their measurement of pollution intensity as a way to assess industry-specific performance in Taiwan against international best practices and over time. They expect their yardstick—pollution intensity per NT$ of value added—to become the metric by which the government judges the environmental behavior of individual firms and industries and its own environmental performance. If this yardstick were tethered to government performance monitoring that linked rewards (e.g., preferential access to subsidized credit or to scarcer and scarcer new land for industrial development), it could spur firms to search for ways to reduce the pollution intensity of output, much the same way it encouraged them to increase exports in an earlier time.[25]

Impact on Environmental Quality

What tangible evidence is there that these myriad plans, policies, and programs actually work? If firm expenditures on pollution control and abatement are any indication, Taiwanese industry seems to have turned an environmental corner. (What follows is from O'Connor 1994, 181–182.) By 1991, four years after the TEPA was created, private-sector investment in pollution control and abatement equipment equaled almost 6% of total private manufacturing investment. In 1992, the year in which direct elections for the president were held for the first time in Taiwan's history, it equaled 4.3% of total investment.

These figures are dwarfed by the pollution control expenditures of state-owned enterprises. In 1992, Taipower allocated 8.4% of its fixed investment to pollution control. Pollution control investment by the China Petroleum Company rose from almost nothing in 1988 to 18.9% of fixed investment in 1990 and to 30% in 1992. By 1992, the total investment in pollution control in state-owned enterprises was nearly three and one-half times larger than in the private sector. Given the large role of state-owned enterprises in Taiwan,[26] it appears that in 1992 industry there expended a larger share of its investment budget on pollution control than Japanese firms did at the height of Japan's pollution control effort.[27]

In addition, many examples suggest that the joint IDB–TEPA Waste Reduction Task Force is providing high-quality technical assistance to firms trying to reduce the pollution intensity of production. Following are several notable examples (IDB n.d., 4–10):

- Factories using stearic acid cadmium as a stabilizing agent have been assisted in a shift to cadmium oxide as raw material and in replacing a highly polluting compound decomposition method with a fusion production process.
- The metal-finishing industry has been assisted in replacing a highly polluting acid wash with airtight sandblasting.
- The scrape metal industry has been assisted in replacing a highly polluting acid wash with a cupric chloride etching solution.
- The electroplating industry has been assisted in a shift to continuous automatic electroplating and to cyanide-free electroplating. It has also been assisted with the installation of equipment to recover chromic acid. The recovery rate is now 98% or more. This kind of assistance has been extended to state-owned enterprises.

- The China Steel Corporation has been assisted in a program to reuse water-quenched clinkers, normally a waste product, from steelmaking. As a result of this program, the annual reuse of clinkers reached 77.5% of output in 1988. In 1993 alone, this reduced dumping of clinkers into the sea by 1.36 million tons.[28]

There is also evidence that this kind of pollution prevention assistance is reducing pollution loads and pollution intensities in a range of industries. Two examples should suffice (IDB n.d., 15). Pollution loads for the major pollutants in the personal computer board industry fell dramatically after emissions controls. The preemissions-control pollution load of lead was 120.6 kilograms a day (kg/d) for firms in this industry. After emissions controls, the pollution load for lead fell to 67.8 kg/d. This met 1993 emissions standards for this industry. The pollution load of chemical oxygen demand (COD) before emissions standards was 32,844 kg/d. This fell to 11,922 kg/d after pollution control devices were installed in firms. Similar success has been recorded in the heavily polluting electroplating industry. Before emissions control, the pollution load of suspended solids equaled 31,504 kg/d; after control, emissions fell to 9,924 kg/d. This was only 13% above the emissions standard for this industry.[29]

Ultimately, if all of these pollution control, abatement, and prevention activities are effective, there should be measurable improvements in ambient environmental indicators. Is there any evidence of this? Several indicators reveal significant and substantial environmental progress. After the first round of environmental regulation in the early 1980s, ambient concentrations of sulfur dioxide and nitrogen dioxide in the air fell steeply (O'Connor 1994, 107). The implementation of a new policy requiring significant reductions in the sulfur content of heavy oil and diesel fuel in 1993 led to even further reductions in sulfur dioxide concentrations.[30] More important, in 1991 16.25% of all days in Taiwan had a pollution standards index greater than 100; by 1994, this fell to 6.99%; and by 1997, to 5.46% (Rock 1996b, 267; TEPA 2000, 1).

Similar improvements are apparent in other ambient environmental indicators in Taiwan. Between 1984 and 1993, particulate matter concentration in the air hovered between 90 and 100 micrograms per cubic meter ($\mu g/m^3$); by 1997 concentration declined to 64 $\mu g/m^3$ for all of Taiwan and 50 $\mu g/m^3$ for Taipei (TEPA 2000, 1). The consumption of chlorofluorocarbons also fell sharply, by 86% (from 16,255 tons in 1988 to 2,304 tons in 1994; Office of Science and Technology Advisors 1995, 31). Although data on lead concentration in the air are not available, the sale of unleaded gasoline rose from about 13% of total sales in 1988 to 65% in 1994 (Office of Science and Technology Advisors 1995, 19). Visible progress has also been made in improving water quality. The unpolluted or slightly polluted portion of major and secondary rivers increased from 46.6% in 1981 to 66.2% in 1999, while the heavily polluted portions of these rivers fell from 14.9% to 12% (Bureau of Water Quality Protection, TEPA 2001).

Conclusions

Why has the Taiwanese government done (and continued to do) all of these things? There are several answers to this question. First and foremost, democratization fostered growing public concern over the environment—as evidenced by a mushrooming environmental protest movement and an almost intractable not-in-my-backyard problem—which meant that the governing elites in the Kuomintang could not ignore industrial pollution.[31] It is not surprising that the KMT learned this the hard way. Following the first steps toward democratization in 1986 and 1987, the government faced an explosion in environmental protests. The major turning point came in 1986, when the government and the powerful economic ministries lost an effort to place a large foreign-owned petrochemical factory in a city that did not want it (Reardon-Anderson 1997; Tang and Tang 1997, 285). After this highly publicized event, known in Taiwan as the "Dupont incident," environmental protests led to the closing of several factories; they also led other factories to shift to cleaner fuels, use less dangerous chemicals, or cut hazardous air emissions (Chan 1993, 41).

By 1987–1988, environmental protests had grown to one a day. It is not surprising that opposition political parties (especially the Democratic Progressive Party) supported local protests, especially when they embarrassed the KMT (Tang and Tang 1997, 285). Criticism from the influential overseas Chinese community, particularly that in the United States, added to these mushrooming domestic pressures.[32] The loss of international recognition has increased elite sensitivity to international criticism. Some of this has extended to the environment.[33] In one instance, the government has gone out of its way to demonstrate exemplary environmental behavior.[34] Growing domestic and international political pressure has been reinforced by basic economic considerations that belatedly have provided an opportunity for the IDB to join in the government's environmental efforts. Four kinds of economic developments predominate.

First, appreciation of the exchange rate, rising wage rates, and emerging labor shortages, along with increased demands for a cleaner environment, contributed to an exporting of industry during the 1980s. Many of the firms that migrated to China and Vietnam, among other places, were in sectors—textiles, leather goods, and metal and electroplating—considered by the Taiwanese to be excessively dirty or polluting. Conversely, the new high-technology industries promoted under the Statute for Upgrading Industries are considered relatively clean. Said another way, the latest competitive shift in industrial structure is away from industries with a high pollution intensity toward those with a low, or at least lower, intensity. Industrial policy, what the IDB does best, simply expedites this process.

Second, there is acute recognition by the government that international competitiveness demands better environmental behavior. Exporters are learning from direct experience that they have to certify to importers in some industrial economies that their products meet importing-economy environmental regulations.[35] Officials in TEPA, CEPD, and IDB are also aware of the competitive value of eco-labeling. For this reason, Taiwan has created its own eco-labeling program: the Green Mark. By 1995, 197 products qualified for Green Mark status (Office of Science and Technology Advisors 1995, 6). These same officials are all too acutely aware of the possible impact of the International Standards Organization (ISO; the standards are known as ISO 14000) on Taiwan's international competitiveness. Because of this, it is not surprising that the government is engaged in bilateral discussions with the European Union to ensure Taiwanese access to the European market following implementation of ISO 14000. It has also established an interministerial committee to consider how Taiwan should respond to ISO 14000.[36]

Third, Taiwanese government officials expect the demand for environmental goods and services in Southeast Asia to grow rapidly during the next several years. They are preparing to capture a substantial share of this market. They see this as part of the next step up the industrial ladder and an important way for Taiwan to distinguish itself in the region. Some, such as IDB officials, see it as a way for Taiwan to start establishing itself as a regional manufacturing center in the Asia-Pacific vertical division of labor (IDB 1995, 27). The National Center on Cleaner Production has contributed to this by setting up an Asian network for clean production.[37]

Fourth, officials of the Taiwanese government view their approach to industrial pollution reduction as more cost-effective than the alternatives. They believe that tough emissions and effluent standards and equally tough enforcement are necessary to get firms to accept that protecting the environment is part of the cost of doing business in a highly competitive global economy. But they believe equally strongly that the IDB's fiscal and financial incentives can quicken the shift to a strategy of less polluting industrial growth. And they know from direct experience that pollution prevention sometimes pays.

For these reasons, then, the Kuomintang and the Taiwanese government have decided to take action and reduce industrial pollution. That action appears to have mobilized not only a relatively new environmental agency but also the preexisting industrial policy machinery, particularly the IDB and CEPD. The CEPD's latest development plan, which promotes the Ten

Emerging Industries, presupposes a structural shift in industry from more to less polluting industries. If this happens, it alone should contribute to a lowering of the pollution intensity of industry. The IDB is complementing this with technical assistance in pollution prevention and research on indicators of pollution intensity per dollar of value added by industry subsector. Taken together, all this activity is quite remarkable. Because it is being done with real data, it puts Taiwan ahead of other economies that are exploring these issues in developing economies.[38] Moreover, when combined with the inspection and enforcement activities of TEPA, it looks as if Taiwan is well on its way to a cleaner environment.

Yet how does Taiwan's pathway to industrial pollution management compare with what happened in Singapore? To begin with, both governments launched their industrial environmental improvement programs in the context of strong technocratic states with substantial embedded autonomy. This meant that there were limited opportunities in both for private-sector actors to oppose the governments' decisions to clean up the environment. In Singapore, promotional privileges for new investments were contingent on promoted firms meeting tough emissions standards. This gave the private sector even fewer opportunities to oppose the government's cleanup efforts. But because the Taiwanese government bypassed industrial policy agencies in its move to clean up industrial pollution, it depended more on TEPA's policing actions and less on the withholding of promotional privileges than did the government of Singapore.

Next, the Taiwanese government was pushed into creating a tough command-and-control environmental regulatory agency by democratization, by the continuing loss of international recognition, and by criticism from an influential overseas Chinese community. This meant that Taiwan's cleanup flowed more from internal and external political pressure than that of Singapore, where the cleanup reflected the actions of a state that faced little such pressure.

Finally, pollution cleanup in Singapore and Taiwan depended on the creation of very traditional command-and-control environmental agencies that were effectively integrated into the institutions of industrial policy. But this integration occurred by different routes in the two economies. In Singapore, it was led by the prime minister, who placed an antipollution unit in his office and thus signaled to industrial policy agencies that they must begin to take the environment seriously. Governing officials in newly democratic Taiwan achieved the same result by initially bypassing the industrial policy agencies and by empowering a more and more competent environmental agency to get tough with polluters. This forced the industrial policy agencies to devise their own approach to pollution management. It also forced them to seek common ground with the regulatory agency and thus contributed to an effective joint TEPA–IDB pollution prevention program.

Notes

[1]This chapter draws on an essay that was initially published as "Toward More Sustainable Development: The Environment and Industrial Policy in Taiwan," *Development Policy Review* 14 (Rock 1996b, 255–272) and subsequently republished in full under the same title in Angel and Rock (2000, 194–208). Permission to use this material has been granted by Basil Blackwell Publishers and Greenleaf Publishing Limited.

[2]Relations between the KMT and U.S. aid advisers also played an important role (Haggard 1990, 83–96).

[3]Taiwan has a relatively large state-owned-enterprise sector, which is concentrated in upstream industry; it accounted for 14% of gross domestic product (GDP) and 33% of gross domestic capital formation in 1978–1980 (Wade 1990, 176). Wade argues that this gives the government substantial control over private industry, which is concentrated in downstream industries.

[4]By the early 1970s, 90% of all farms were less than two hectares, and 80% of the agricultural population were owner-cultivators (Amsden 1979, 349).

[5]Between 1960 and 1982, the statute was amended 11 times (Wade 1990, 183).

[6]Total emissions of sulfur oxides, particulates, and nitrogen oxides in Japan as a share of GDP are less than one-quarter of the OECD averages (World Bank 1992b, 40).

[7]The overseas Chinese community has had a long-standing reciprocal relationship with the Nationalist government that stretches back to when it controlled mainland China. This pattern continued once the Nationalist government moved to Taiwan in 1949. Since then, the Nationalist government has viewed overseas Chinese, including overseas American Chinese, as part of China and Taiwan (Lung-chu and Lasswell 1967, 246). This reflects, among other things, Taiwan's claim that it is the only legitimate government of China. In addition, the overseas Chinese American community has been particularly important to the Nationalist government in China and in Taiwan. Before the collapse of the Nationalist government on the mainland, this community worked hard with the so-called China lobby in the United States to increase aid to the Nationalists (Melby 1968, 26; Chern 1976–1977, 634–635). Following the KMT's move to Taiwan, the Nationalist government invested heavily in maintaining influence with the U.S. government through the China lobby. There is substantial evidence that before U.S. normalization of relations with the People's Republic of China, this community was enormously successful at influencing U.S. foreign policy (Seagrave 1985, 432–446). Thus, it is not surprising that the government has continued to cultivate the overseas Chinese community, particularly in the United States. Nor is it surprising that the government of Taiwan has been sensitive to criticism from the overseas Chinese community.

[8]Amendments tightened biological oxygen demand (BOD) and chemical oxygen demand (COD) emissions standards, granted the TEPA authority to levy a water pollution fee on water users, and mandated that the government's water-quality-monitoring results be made part of the public record (O'Connor 1994, 68).

[9]Amendments to the Air Pollution Control Act tightened emissions standards for sulfur oxides and carbon monoxide and granted local governments authority to set standards that were stricter than the national standards (O'Connor 1994, 68).

[10]The government also engaged the U.S. EPA to provide technical assistance to the TEPA (interviews in the United States and Taiwan, November 1995).

[11]For air, this is true for sulfur dioxide, carbon monoxide, nitrogen dioxide, lead, and photochemical oxidants. Taiwan's ambient standard for total suspended particulates is lower than that for the United States. For water, Taiwan's standards for chemical concentrations for cadmium, cyanide, chromium, arsenic, and mercury are equal to or more stringent than U.S. standards (O'Connor 1994, 113–115).

[12]The setting of emissions standards begins with an expert committee composed of academics, government officials, and others who advise on standards. The process permits informal consultations with industry, but this is not required (O'Connor 1994, 91).

[13]This appears to be one of the ways the government is communicating its intentions to get tough on polluters.

[14]The process also seems to be changing. Representatives of the National Federation of Industries (NFI), the premier industry association in Taiwan, state that members from their Pollution Control and Industrial Safety Committee go over proposed TEPA regulations line by line. This is followed by attendance at TEPA public hearings for all new regulations. In one recent instance, the NFI got the TEPA to water down an emissions standard for BOD because firms in the porcelain industry said they could not meet the tougher standard (interviews in Taiwan, November 1995).

[15]In the words of one academic interviewed, tough standards are necessary to get the large number of small and medium-sized enterprises to understand that pollution control is now part of the cost of doing business (interviews in Taiwan, November 1995). This also appears necessary to get bigger enterprises and industrial policy agencies such as the IDB to understand that the government is serious about reducing industrial pollution.

[16]Between 1990 and 1994, an average of 37 factories was closed annually (Office of Science and Technology Advisors 1995, 29).

[17]Government laws require firms to organize into an industry association whenever there are more than five firms in an industry (interviews in Taiwan, November 1995).

[18]For example, the government is planning to build 22 large waste incinerators and to contract construction out to private-sector firms. Private engineering firms in the Taiwan Environmental Engineering Association expect to get most, if not all, of this business (interviews with members of this industry association in Taiwan, November 1995).

[19]Because of rampant NIMBY (not in my backyard) problems, preferential access to new industrial land is extremely important.

[20]The tax credit for imported equipment is only 10%.

[21]In 1992, domestic production of the environmental goods and services industry equaled $1.3 billion. Of this, 6% ($75 million) was exported. Production is expected to grow at an average annual rate of 11% through 2002. In that year, domestic production will equal $3.75 billion, of which $464 million is expected to be exports (14%) (IDB n.d., table entitled "Status and Target of Environmental Protection Industry").

[22]A representative of a U.S. firm trying to break into this business stated that his firm had a difficult time doing this because the government's assistance was both very good and free (interviews in Taiwan, November 1995).

[23]This is part of UCL's Environmental Sciences and Technology Division. The division works on ISO 14000 and life cycle analysis, and it houses the new National

Center on Cleaner Production (interviews at UCL of ITRI in Taiwan, November 1995).

[24]The IDB's administration of the duty-drawback system requires about 20 technicians working full time to calculate the input–output coefficients for large numbers of export items and the imported inputs used to produce those items (Wade 1990, 140). Because many of the inputs in production are imported, it appears that it might be relatively easy to adapt this scheme to calculate the pollution intensity of value added by industry subsector.

[25]If this happened, it could well serve as the basis for more cost-effective environmental outcomes. For a discussion of this in another context, see Porter and van der Linde (1995, 110–114).

[26]Taiwan's relatively large state-owned-enterprise sector accounted for 14% of gross domestic product and 33% of gross domestic capital formation in 1978–1980 (Wade 1990, 176.)

[27]This occurred in 1975 in Japan. In that year, Japan allocated 7.5% of the fixed investment budget of industry to pollution control (O'Connor 1994, 181).

[28]In that year, only 22.5% of clinker waste (540,000 tons) was dumped into the sea (TEPA 1993, 328).

[29]Comparable numbers for other pollutants include 5,650 kg/d of nickel before emissions control and 633 kg/d after control; 5,996 kg/d of chromium before control and 495 kg/d after control; 5,403 kg/d of zinc before control and 1,253 kg/d after control; 4,382 kg/d of copper before control and 758 kg/d after control; and 2,746 kg/d of iron before control and 195 kg/d after control (IDB n.d., 15).

[30]Following this, the sulfur dioxide concentration in the air (in parts per billion) fell from 26 in 1993 to 8 in 1994 (Office of Science and Technology Advisors 1995, 17).

[31]Recent public opinion polls put industrial pollution as one of the top three problems facing the country. The NIMBY problem has made industrial siting almost impossible. One response to this has been to dredge the ocean to create new offshore islands on which the most polluting industries will be located (interviews with IDB officials in Taiwan, November 1995).

[32]Each year, the government holds a national reconstruction conference, at which scholars, government officials, and overseas Chinese are invited to review accomplishments and assess the major issues facing the country. At least since the mid-1970s, the overseas Chinese, particularly those from the United States, have been critical of Taiwan's growing pollution problems (interviews with CEPD officials in Taiwan, November 1995).

[33]Condemnation by the United States of Taiwan for violating the Convention on International Trade in Endangered Species has stung the country's political elite (interviews with TEPA officials in Taiwan, November 1995).

[34]Despite the fact that Taiwan was not permitted to be a signatory to the Montreal Protocol, it met the protocol's stipulations (Office of Science and Technology Advisors 1995, 31).

[35]Wood furniture manufacturers who export to the Nordic countries must certify that the wood in furniture does not come from virgin tropical forests (interviews with IDB officials in Taiwan, November 1995).

[36]In addition, the private sector is looking to the government to advise it on what ISO 14000 will mean for them (interview with representative of the NFI, November 1995).

[37]This is being done in cooperation with the Federation of Asian Chemical Societies, Hong Kong Productivity Council, Philippine Business for the Environment, Korean Institute for Chemical Technology, and Asian Institute for Technology in Thailand (interviews with ITRI officials in Taiwan, November 1995).

[38]The World Bank is extrapolating U.S. pollution intensities of specific industries to developing countries. By projecting changes in industrial structure, it is able to project changes in the pollution intensity of industry (World Bank 1994a, 77–81).

4

Searching for Creative Solutions to Pollution in Indonesia

The conditions that led to more effective industrial pollution management in Singapore and Taiwan appear to have been largely absent from New Order Indonesia. Before the collapse of the New Order government, intense international pressure on the Indonesian government from the donor community and nongovernmental organizations (NGOs) to reform its environmentally destructive forest policies fell on deaf ears.[1] The same might be said of international pressure to improve industrial pollution management.[2] A deep-seated mistrust of markets and foreign investment—alongside a large domestic market that made it possible for the government to sustain its promotion of import-substitution industries, even while it was promoting exports—appears to have limited the effect of international environmental market pressure on domestic producers.[3] Indonesia has a low per capita income; a fairly small industrial sector (until recently); a relatively small educated urban middle class; and an authoritarian government that controls the media and independent organizations in civil society. All of these factors contributed to the lack of public pressure to clean up the environment (World Bank 1994a, 173).[4] And each factor was reinforced by pervasive patrimonial networks between high-ranking government and military officials and big business that made it difficult for the New Order government to achieve a consensus on pollution management.[5]

This combination of forces at least partially explains why Indonesia's environmental impact management agency, BAPEDAL, remains weak, understaffed, and overcentralized in Jakarta. It also helps to explain why BAPEDAL lacks the resources, technical capacity, and legal authority to mount an effective nationwide inspection and enforcement program (World Bank 1994a). And it explains why there has not yet been a significant, sustained improvement in environmental quality in Indonesia. Despite these shortcomings, BAPEDAL, its parent agency (the State Ministry for Population and the Environment, or SMPE), and several local public-sector agencies (e.g., BAPEDALDA SEMARANG) have been able to design and implement several unique, successful (though partial and highly targeted) industrial pollution management programs.

In 1989, the SMPE launched a highly innovative and effective voluntary industrial wastewater emissions reduction program, PROKASIH (or Clean Rivers), which brought about large declines in biological oxygen demand (BOD) loads and in the BOD intensity of PROKASIH plants (Afsah et al. 1995, 23–24). In 1993, BAPEDAL followed up with an even more innovative and effective public disclosure program, PROPER PROKASIH, which until recently rated and disclosed the environmental performance of nearly 200 of the country's largest factories and most significant industrial water polluters (Afsah and Vincent 2000). It also appears to have worked, at least for a while.[6] At the local level, following a highly publicized pollution incident, the mayor of one of the country's largest cities, Semarang, launched a partially effective small-scale monitoring and enforcement program, despite the fact that he did not have the central government's permission (Aden and Rock 1999).

Numerous aspects of these partial, targeted, and program-based approaches to industrial pollution control are particularly intriguing. These approaches:

- worked—at least for a while;[7]
- conserved scarce regulatory resources and limited regulatory capabilities;
- did not depend on the creation of a comprehensive pollution control agency with the technical capacity and full legal authority to mount a rigorous national inspection and enforcement program—and consequently were less dependent on the country's bifurcated state;
- were based on pragmatic, trial-and-error approaches to industrial pollution management;
- appear to have been relatively free from corruption;[8]
- had the support of the president;[9]
- engaged others in addition to the government—neighbors, communities, the media, environmental NGOs, and polluting plants—in pollution management;
- drew on new policy ideas emanating from the Organisation for Economic Co-operation and Development (this applies only to some, particularly Clean Rivers and PROPER PROKASIH);

- are being copied by pollution management agencies outside Indonesia;[10] and
- stand in marked contrast to the almost complete absence of successful instances of industrial pollution management outside these programs within Indonesia,[11] in more democratic and prosperous Thailand, and in the more democratic Philippines.

Evolution of Pollution Management

How and why have BAPEDAL and local environmental managers been able to design and implement such innovative, effective (albeit partial), targeted, and program-based approaches to managing industrial pollution? This is an intriguing question, especially because the New Order government did not appear to be either very committed or able to implement credible, comprehensive industrial pollution control or forest policies. Answers to this question can be found in the history of the government's early failed efforts to control pollution. They can also be found in the broader political economy of policymaking in Indonesia. Before turning to the influence of the latter, it is essential to examine the government's early efforts to manage pollution.

Early Efforts

Government concern for the environment began with preparation for the 1972 United Nations Conference on the Human Environment in Stockholm (Cribb 1990, 1125–1126). This was followed by the creation of a high-level government committee, chaired by the vice chair of BAPPENAS, the country's powerful economic planning agency, which was charged with advising the government on how it might integrate environmental considerations into its development plans, policies, and programs (World Bank 1994a). By this action and by including a policy statement in BAPPENAS' next five-year plan (the third, commencing in 1978) to the effect that the environment should be protected from undue damage, the government sent the first signals of its intention to start protecting the environment. Also in 1978, the government established a State Ministry for Development Supervision and the Environment (SMDSE), and the Ministry of Industry (MOI) directed manufacturers to avoid and overcome the pollution associated with their activities.[12] As Cribb (1990) states, the purpose of the MOI's directive was to signal to private firms that it also was getting serious about cleaning up the environment.

In 1982, the government passed landmark environmental legislation, the Act Concerning Basic Provisions for the Management of the Living Environment. This act institutionalized requirements for environmental impact assessments (EIAs), empowered the SMDSE to coordinate environmental policy, granted provincial governors executive power over environmental matters, opened the way for the development of new quality standards to protect the environment, and explicitly assigned responsibility for cleaning up pollution to polluters (O'Connor 1994, 71).[13] In 1983, the government created the cabinet-level SMPE and assigned a highly respected economic technocrat to lead it. In 1986, the SMPE strengthened the legal requirements for EIAs by announcing a set of regulations governing their implementation (World Bank 1994a, 185).

These early formal actions by the central government were reinforced by a series of local actions that suggested the central government could no longer afford to ignore the environmental consequences of rapid industrial growth (Cribb 1990, 1128–1130). In 1977, the provincial government of East Java attempted to reduce the pollution of the Surabaya River by announcing that no new construction permits would be issued to manufacturing plants locating along the river. In 1980, farmers in a Jakarta suburb burned down a state-owned fertilizer factory because its wastewater emissions were polluting their irrigation channels. In 1986, another local government found a paper factory responsible for polluting the Bekasi River, and the factory was ordered to treat its effluent. In 1988, the deputy governor of East Java pushed some of the worst polluters along the Brantas River to sign agreements to reduce their pollution or face legal action. Similar developments occurred elsewhere in Jakarta and in Sumatra.

An important question remained, however: What could the government do to slow the deterioration of urban environments resulting from high-speed industrial growth? Before the creation of BAPEDAL in 1990, the government settled on a three-pronged strategy to improve industrial pollution management. First, because the SMPE lacked the implementing authority, budget, and staff to mount a full-scale inspection and enforcement program, it turned to EIAs as a way to get line agencies to integrate environmental considerations into their everyday activities.[14] Although EIAs had been mandated in Indonesia's landmark environmental act of 1982, little implementation had been done. This was corrected in 1986, when the SMPE announced the implementing regulations for EIAs.[15]

Second, because Indonesia lacked sufficient expertise on environmental issues, the SMPE invested in the development of indigenous technical capacity in environmental sciences and management by establishing and financing 54 environmental study centers (PSLs) at universities in each of Indonesia's 27 provinces (World Bank 1994a, 182). The SMPE expected that PSL staffs and those they trained would serve on EIA teams and provide policy advice to the government. Third, to create more public pressure for environmental cleanup and to enhance the ability of the public to participate in EIAs, the SMPE also fostered the development of environmental NGOs, environmental awareness, and environmental education. This included increasing the awareness of the press. Given the New Order government's antagonism toward even quasi-independent groups in civil society and a free press, this support was quite remarkable.

Unfortunately, weaknesses in each element of this three-pronged strategy ultimately undermined its effectiveness. The development of a cadre of well-trained environmental professionals in the PSLs was slow and highly uneven. By 1994, fewer than half of the 54 PSLs had good reputations, and only 12 were playing a significant role in either EIAs or in offering policy advice at any level of government (World Bank 1994a, 182).

A different set of problems plagued public participation in EIAs. Because the New Order government both feared and controlled organized groups in civil society, most government agencies were simply too reluctant to include environmental NGOs or the public in discussions of their projects. Weak participation by newly trained environmental professionals in line-agency EIAs, and limited or nonexistent participation by NGOs and the public in those EIAs, meant that the burden of EIAs fell on the implementing agencies. In the end, this burden was too heavy to bear. Because most line agencies viewed EIAs as superfluous add-ons, they failed to integrate them into preproject feasibility studies and analyses of proposed projects. Moreover, line agencies frequently faced conflict between those wanting to push a particular project forward and those wanting to use an EIA to rigorously screen the project for its environmental effects (World Bank 1994a, 270–271). It is not surprising that the former tended to win over the latter.

The Innovative PROKASIH Program

The disappointing experience with its EIA-based environmental strategy ultimately led SMPE officials to look for alternatives. This led in 1989 to the creation of an innovative, highly targeted pollution control program: PROKASIH, or Clean Rivers. PROKASIH focused on BOD in industrial wastewater emissions; it was based in the signing of voluntary, but not legally binding, pollution reduction agreements between major industrial BOD water polluters, vice governors of provinces, and the SMPE. (The next two paragraphs, unless noted, are based on Afsah et al. 1995.)

Given the limited resources and limited implementation capabilities of the SMPE, PROKASIH was forced to draw on the strength of local government, particularly provincial vice governors. BAPEDAL, which was created in 1990, invited provinces to participate in PROKASIH, and each participating province was asked to assemble a local PROKASIH team headed by vice governors and made up of local public officials and representatives from PSLs and testing laboratories. Once constituted, PROKASIH teams were responsible for identifying the most polluted rivers, or most polluted sections of rivers, within a province and the major industrial water polluters along those rivers. Once identified, vice governors invited major BOD water polluters, such as pulp and paper plants and food processing plants, to sign a voluntary, legally nonbinding pollution reduction agreement with the vice governor and the SMPE. Then the BOD polluter and the PROKASIH team took repeated samples of a plant's effluent to establish a pollution baseline and to assess the degree to which the industrial facility was meeting the terms of the agreement.

PROKASIH teams were funded jointly by the SMPE and provincial governments. Initially, 8 provinces agreed to participate in the program. Participation rose to 13 provinces in 1994 and 17 in 1996. By the end of 1994, when the World Bank conducted an evaluation of the program, about 1,400 manufacturing plants had participated. The available evidence suggests that the program was highly successful. The mean reduction in BOD loads from PROKASIH plants between 1990 and 1994 was 44%. The BOD load per unit of output fell by 55%. In addition, BAPEDAL learned that a small number of large BOD polluters were responsible for most of the industrial BOD load in particular rivers.[16] This gave BAPEDAL important information to use in mounting an even more targeted pollution reduction program.

Why did PROKASIH work so well? There appear to be three answers. First, given the strength of the Indonesian government and the weakness of business (especially Sino-Indonesian business), manufacturing plants (large plants in particular) apparently felt they could not reject a request from vice governors to participate, nor could they easily ignore their agreements once they were signed.[17] Second, the program took advantage of the fact that some plants in Indonesia were abating BOD pollution, sometimes quite considerably, despite the government's lack of legal authority to inspect plants and enforce emissions standards. Third, the minister of the SMPE was willing to publicly embarrass those facilities that failed to meet their pollution reduction agreements.[18] This combination suggests that PROKASIH allowed the government to take advantage of what might be called reputational regulation.[19]

The Creation of BAPEDAL

Shortly after launching PROKASIH, the government and the SMPE set about creating a traditional command-and-control environmental agency (O'Connor 1994, 85, 91). This commenced with the creation in 1990 of an environmental impact management agency, BAPEDAL, to ensure the effective implementation of EIAs and develop and enforce environmental standards. In 1991, BAPEDAL set effluent and emissions standards for water for 14 industries, including caustic soda production, electroplating, leather tanning, petrochemicals, pulp and paper, textiles, rubber and sugar production, fertilizers, and plywood manufacturing. Standards were set after gathering evidence on best-practice technologies used in Asia, discussing the setting of standards with relevant industry associations, and reaching agreement with those associations and other government ministries. In 1992, the SMPE tightened emissions standards for several air pollutants, including sulfur dioxide, carbon monoxide, nitrogen dioxide, ozone, and total suspended particulates. BAPEDAL also took over the management of PROKASIH.

Along with these developments, the government took a number of steps to enhance the technical capabilities of its new command-and-control environmental agency. Despite an effort to reduce the overall size of government, both BAPPENAS and the Ministry of Finance supported building BAPEDAL into a competent command-and-control agency by assigning a top-echelon government official to it. Subsequently, BAPEDAL's staff, which had started at about 30, rose to 100 by 1994 and 300 by 1996 (interviews with a World Bank adviser to BAPEDAL, 1996).

As a part of this effort to grow BAPEDAL, the government signed a number of loan and grant agreements with various aid donors. The U.S. Agency for International Development supported a clean production and pollution prevention project run by BAPEDAL. The Japanese initiated a soft-loan program for the purchase of pollution control equipment by Indonesian firms (interview with the program leader of this project, 1996). They also launched a project in BAPEDAL's Environmental Management Center, which was designed to enhance the agency's capacity to gather and analyze data on ambient air and water quality, as well as emissions data from stationary sources (interview with a World Bank adviser to BAPEDAL, 1996). The World Bank financed several capacity-building projects in BAPEDAL. One focused on developing rigorous protocols for standard setting, monitoring, enforcement, and compliance management. Another worked on developing the legal mandate for BAPEDAL to do these things. A third explored establishment of local (provincial) environmental impact agencies, known as BAPEDALDA.

Despite these developments, BAPEDAL continued to lack the capacity to reliably monitor ambient air and water quality and the legal authority to inspect and enforce its tougher and tougher air and water emissions standards. As a consequence, air and water monitoring was limited, irregular, and unreliable (World Bank 1994a, 174–175). This reflected, among other things, poor quality control for data collection and analysis, outdated and inadequate equipment, a lack of standard protocols, and a shortage of well-trained staff. Monitoring and enforcement of emissions standards were even weaker and less reliable. Monitoring of water emissions, which was assigned to provincial governors, was mostly haphazard. This was due in part to the fact that only the MOI and local police had the authority to enter factories to take emissions samples. The MOI rarely did so, whereas the police only did so in conjunction with BAPEDAL's public complaints program.[20] To make matters worse, BAPEDAL did not have the authority to issue permits detailing emissions requirements, and the courts refused to grant legal standing to either BAPEDAL's emissions standards or to the results of monitoring. This made it virtually impossible to prosecute those who violated standards. It is interesting, however, that this did not stop the country's new environmental NGOs from suing polluters and the government for failing to live up to mandated emissions and quality standards.[21]

More Alternatives and PROPER PROKASIH

Faced with these problems, local governments, owner-operators in a number of industrial estates, and senior officials of BAPEDAL began searching for other ways to ameliorate pressing industrial pollution problems. After a highly publicized pollution incident in the city of Semarang (the country's fourth largest city, on the north coast of Java) attracted the attention of Indonesia's Legal Aid Society, the city's government created its own industrial pollution management agency, in a joint effort with WALHI (the country's umbrella environmental NGO) and BAPEDAL.[22] This new agency, with the backing of the mayor, launched a small-scale monitoring and enforcement program.

A recent empirical study of the effectiveness of Semarang's program revealed that it had three effects on plants. First, it increased the probability that plants would install pollution control equipment. Second, it led plants to undertake environmental audits, adopt pollution reduction goals, and create environmental management positions. And third, it significantly increased plants' environmental protection activities (Aden and Rock 1999, 367). At about the same time, the owners of about 20 industrial estates on Java began to install central wastewater treatment facilities for plants located on their estates. The owners used wastewater emissions charges to fully recover the costs of treatment, and they required some facilities to pretreat their emissions before sending them to the central facilities.[23] More recently, the provincial government of East Java has considered charging effluent discharge fees to reduce industrial pollution flowing into the Surabaya River (Rock 2000c, 16).

For its part, BAPEDAL turned its attention to building on the success of PROKASIH by turning it into an environmental rating and public disclosure program called PROPER PROKASIH.[24] The rating system took advantage of the fact that the actual compliance status of PROKASIH plants varied considerably. Some PROKASIH plants went beyond complying with emissions standards, some just met the standards, some made an effort but fell short of compliance, and some made no effort. The senior staff of BAPEDAL reasoned that publicly disclosing these variations might end up rewarding those at or exceeding compliance and shaming those with poor ratings into doing better (Afsah and Vincent 2000, 161).

Although the idea behind PROPER PROKASIH was conceptually simple and potentially powerful, as far as the senior staff of BAPEDAL knew, no environmental agency had ever attempted to improve a plant's environmental compliance by rating, ranking, and disclosing its performance.[25] Moreover, the BAPEDAL staff soon realized that implementing PROPER PROKASIH would require overcoming a significant number of problems. (The rest of this and the next few paragraphs, unless noted, are drawn from Afsah and Vincent 2000.) To begin with, BAPEDAL had to develop a simple, reliable, transparent, and honest rating system that would easily convey a plant's compliance status to the public and the media. Because of extensive patrimonial links between high-ranking government and military officials and the big Sino-Indonesian conglomerates, BAPEDAL staff recognized that they needed high-level government support for PROPER PROKASIH, including that of the president. They also recognized that the program's credibility rested on their ability to correctly classify a facility's compliance status. They knew that environmental NGOs and community leaders would criticize them if they were too easy on facilities, but they also knew that owners of facilities (and their government and military supporters) would criticize them if they were too harsh.

PROPER PROKASIH's senior managers ultimately solved each of these problems. Given the importance of BOD loads to PROKASIH and to the water quality of Indonesia's rivers, they focused their rating system on the government's relatively new national regulations for BOD emissions for 14 different industries. The emissions standards in these regulations had three advantages:

- They provided BAPEDAL with a simple focus.
- They identified a major industrial pollutant of the country's rivers.
- This pollutant was the only one with which BAPEDAL had experience.

Because these BOD emissions standards were developed through an extensive consensus-building and consultative process with industry, environmental NGOs, and other government agencies, they were widely acknowledged and accepted within Indonesia.

Once the decision had been made to focus on BOD emissions, BAPEDAL began to develop a simple rating system, build political support for the program, and formulate protocols to minimize the risks of incorrectly rating a plant. BAPEDAL's senior staff ultimately settled on a simple rating system that assigned color codes with distinct cultural meanings in Indonesia to each compliance status (Afsah and Vincent 2000, 161). Manufacturing plants that made no effort to meet BOD emissions standards were to be rated black. Those that made some effort, but not enough, to meet standards were to be rated red. Those that just met standards were to be rated blue.

Those that exceeded compliance with standards were to be rated green, and those with a compliance status at international best-practice levels were to be rated gold.

BAPEDAL secured the political support of the president and the minister of the SMPE by proposing appointment of a special advisory board (with representatives from BAPEDAL, the Department of Health, industry and business associations, and environmental NGOs) to review the assignment of ratings that might be politically problematic because they were very high or low (gold, green, or black). BAPEDAL also attempted to minimize the risk of error by using multiple emissions samples that were independently obtained and analyzed and by designing a simple yes-or-no, computer-based system to assign the color-coded ratings to particular facilities.[26] To ensure that participating plants, the media, and NGOs understood how the system worked, representatives of each were given demonstrations.

Because BAPEDAL had existing relationships with PROKASIH plants, the first set of ratings was limited to many of these plants and several others.[27] The available evidence suggests that PROPER PROKASIH was quite successful (Afsah and Vincent 2000, 168–169). Of the 187 facilities that were initially rated in June 1995, none received a gold rating; 5, a green rating; 61, a blue rating; 115, a red rating; and 6, a black rating. BAPEDAL rewarded the 5 green, beyond-compliance plants by announcing their names and publicly commending them for their exemplary performance. To protect itself from incorrect ratings, BAPEDAL did not identify those plants that received blue, red, or black ratings. Instead, it simply announced the number of plants in each category, consulted with these plants to ensure that their ratings were correct, and informed them that they had six months to improve their performance. After that, BAPEDAL said it would re-rate each plant and announce the results.

Six months later, BAPEDAL released the results of its reratings. There were 4 (20% fewer) green plants, 72 blue plants (18% more), 108 red plants (6% fewer), and 3 black plants (50% fewer). This shift among rating categories meant that the compliance rate had increased from 36% to 41%. Because BAPEDAL worked with plants receiving red and black ratings, implemented a simple and tamper-free rating system, and secured and maintained high-level political support for the program, not one plant protested its rating. Protest from community leaders and NGOs was muted.[28]

Why has this program, like PROKASIH, worked so well? There are six reasons. First, BAPEDAL's senior managers gained and kept high-level political support for the program. Without this, and especially without the president's support, it would not have been possible to rate facilities red or black. Second, they reached out to manufacturing facilities, environmental NGOs, community leaders, and the media to demonstrate how the ratings worked. This inspired confidence in the program among all these actors. Third, because its managers focused the program on widely accepted national BOD emissions standards, there were limited grounds for criticism. Fourth, they grafted PROPER PROKASIH onto the well-recognized, successful PROKASIH. Fifth, they built on what they had learned in PROKASIH. This suggests that the designers of PROPER were exceedingly pragmatic. Sixth, they kept the rating system simple, credible, transparent, and honest. Without this, no one would have accepted the results of PROPER.

The Politics of Policymaking and Pollution Management

The picture painted above suggests that BAPEDAL's senior officials, as well as those of several local government agencies, have been pragmatic, creative, committed, and capable of devising practical solutions to the pressing industrial pollution problems the country faces. It also suggests that they have been able to get results within the confines of an otherwise weak national inspection and enforcement capacity within BAPEDAL and in Indonesia's provinces. Although potentially compelling, this story suffers from what some might consider a fatal flaw: It appears to be largely inconsistent with reigning conventional wisdom about Indonesia's rent-seeking and bifurcated state—a state that has been seen as quite good at macroeconomic policy (monetary, fiscal, debt, and exchange rate policy) but quite ineffective at microeconomic policy.[29] Because industrial pollution control is microeconomic policy par excellence, it should have been subject to rent-seeking, which would have prevented BAPEDAL and local governments from successfully implementing it.

Apparently that has not been the case. Why? It is undoubtedly true that those in the agencies dealing with microeconomic policy, particularly in the Ministry of Industry and Trade, opposed the development of a more effective national and provincial program for inspection and enforcement.[30] As a result of this opposition, government officials who were interested in improving pollution management were forced to be creative and innovative. What is most surprising, in light of the conventional wisdom about the New Order state, is that they have been relatively successful in devising and implementing innovative pollution programs that even the president has supported.

How and why has this happened? Fragmentary evidence gained in interviews suggests that senior officials of the New Order government, including the president, were slowly coming to see improved industrial pollution control in their own as well as the nation's interest. This is manifest in the president's support for both PROKASIH and PROPER PROKASIH. It is extremely doubtful that he would have lent his support to both had they not been in line with his *and* the nation's interests.[31] The same might also be said about the growing interest of the macroeconomic technocrats in pollution control policy. It is unlikely that officials of the Ministry of Finance and of BAPPENAS would have agreed to "grow" BAPEDAL into an effective command-and-control agency if they thought pollution was unimportant, particularly because this was occurring when both were trying to deregulate the economy and reduce the overall size of government employment. It is equally unlikely that a respected senior macroeconomic technocrat would have agreed to become the minister of the SMPE unless he also thought it was important.

What might have led the president and at least some of the macroeconomic technocrats to conclude that it was time for Indonesia to pay more attention to industrial pollution? Pressures from donors and international NGOs, even if they were largely ignored, had some influence. Discussions in international forums—such as the Brundtland Commission and the United Nations Conference on Environment and Development in Rio de Janeiro—probably had an impact. A deteriorating urban environment, which was plain for all to see, had some effect on government macroeconomists, who were undoubtedly also influenced by the growing literature on the human health costs of environmental degradation (World Bank 1994a, 88–90). But by themselves, none of these was likely sufficient to get the New Order government to permit nascent national and local environmental agencies the discretion to develop and implement a range of innovative pollution control policies.

The technocrats and President Soeharto were most likely influenced by the rapid growth and spread of public protests against polluters after 1980. As Cribb (1990) has demonstrated, from the early 1980s there was a growing number of unorganized public protests against a deteriorating environment. Sometimes, the protests led to uncontrolled violence, as in the burning of a state-owned fertilizer factory by farmers. Sometimes protests resulted in legal attacks on the New Order government by Indonesia's Legal Aid Society. Sometimes, they contributed to consumer boycotts organized by WALHI (World Bank 1994a, 202). And they often contributed to a consistent drumbeat of negative media reporting on the environment. The New Order government has always been sensitive to internal and external threats to its own and the national interest. When such threats have been perceived, the government has, at least until the financial crisis that helped topple the regime, been able to act decisively. This has been particularly true with macroeconomic policy—in 1965 and 1966, during the Pertimina crisis, and again in the mid-1980s following a slowdown in economic growth.[32]

What is less well understood, however, is that when it counted, the New Order government has also been able to overcome its patrimonial distributional networks and a bifurcated state to make hard microeconomic policy choices. This is most clear in rice agriculture, for which the macroeconomic technocrats in the Ministry of Finance and BAPPENAS worked closely with those in microeconomic agencies to promote rice self-sufficiency (Bresnan 1993; Timmer 1975, 1989, 1993, 1996). We know that Soeharto supported this drive for self-sufficiency because he saw it in his and the national interest (Glassburner 1978a, 143). We know he urged the technocrats to be more pragmatic and to pay more attention to how rice policies were designed and carried out (Bresnan 1993, 127). We also know that this effort required substantial intervention in markets for inputs, in credit markets, and in output markets. In each instance, the government and the technocrats in macroeconomic agencies deliberately distorted market prices. The objective of intervention in output markets was to stabilize the domestic price of rice around its world price. Achieving this required substantial coordination across several macroeconomic and microeconomic agencies, including Bulog, the food logistics agency; BAPPENAS; the Ministry of Finance; the Ministry of Agriculture; the president's office; and the Coordinating Ministry for Economics, Finance, and Industry. This gave one of the key macroeconomic agencies substantial influence over one of the key sectoral microeconomic agencies: Bulog. Although this effort has not been without its own problems, overall it has been quite successful.[33]

Yet this was not the only intervention in rice markets organized by the macroeconomic technocrats. Because farmers had limited experience with commercial fertilizers, pesticides, and high-yielding seeds, the government subsidized each, and it also subsidized credit to farmers so they could purchase these new inputs.[34] These subsidies, along with an aggressive, publicly funded agricultural extension program (Hill 1996a, 129), were used to overcome failures in information markets by significantly improving the price ratios of output to inputs (Booth 1989, 1243). Much of this was funded out of the revenue windfall that accompanied the oil price shocks of the 1970s.[35] Once information failures were overcome, subsidies were removed. And once the government recognized the negative environmental consequences of pesticide input subsidies, it removed them and mounted an effective, integrated pest management program (World Bank 1994a, 198).

This example suggests that the conventional view of the New Order state as bifurcated and fraught with rent-seeking patrimonial distributional networks is at least somewhat overdrawn. This does not mean that the conventional view is completely wrong. It is not. There has been some bifurcation between macroeconomic and microeconomic policymaking in the New Order state. But when the national interest dictated overcoming this bifurcation (as in the case of rice), it was subdued. The same can be said about rent-seeking patrimonial distributional networks between high-ranking government and military officials and big Sino-Indonesian businesses. Those networks did and do exist. They did and still do foster corrupt and rent-seeking policies. But despite this, the New Order government has, at least until the recent financial crisis, been able to overcome these networks to make hard microeconomic policy decisions. It did this not only in rice agriculture and in integrated pest management, but also (as has been argued elsewhere) in industrial policy, where most have only seen rent-seeking policies.[36]

Except for macroeconomic policy, the most successful policy interventions have required coordination between macroeconomic technocrats in central agencies and those who control the relevant microeconomic agencies. In these instances, success has depended on the pragmatic identification and implementation (through trial and error) of workable interventions. Once this is recognized, BAPEDAL's PROKASIH and PROPER PROKASIH (the small-scale monitoring and enforcement program initiated by the mayor of Semarang) programs and the other partial and targeted approaches to pollution management described here are all quite explicable. They look very much like what the New Order government has almost always been able to do.

Conclusions

Because Indonesia's level of development and political economy most resemble those of other poor economies in Asia and Sub-Saharan Africa, it is important to ask what lessons the governments of these economies might learn from the experiences of Indonesia. Several seem particularly relevant. Successful institutional innovation in Indonesia has depended on creating at least minimum levels of technical capability in a national environmental agency. It has also depended on attracting the right people—those who are committed to making a difference and those who can draw on their strong links to the country's top leadership for support. Without both, it is doubtful that much would have happened. Environmental policy innovators also have been politically and institutionally shrewd. They have timed their interventions to highly publicized pollution incidents. They have learned how to husband scarce regulatory resources and limited technical capabilities by focusing on particularly pressing environmental problems. They have developed clear, simple, credible, honest, and transparent ways of operating. They have reached out to NGOs, communities, the media, and of course polluters. And they also have obtained and retained the support of the president for many programs.

Two other lessons emerge from Indonesia's experience. First, there is no evidence that any of these programs has led to improvements in ambient environmental quality. Second, virtually none of these programs has survived the recent financial crisis and the toppling of the New Order government. These lessons suggest that lasting, sustainable industrial pollution control programs that contribute to the improvement of ambient environmental quality most assuredly depend on creating tough command-and-control environmental agencies that can weather political change. What this means is that Indonesia will need to find a better balance between its innovative, information-based programs and traditional regulatory programs. Chapter 7 describes what such a program might look like.

Notes

[1]Numerous studies document the effect of rent-seeking policies on forest management (Barr 1998; Repetto 1988; World Bank 1989, 1994a; Sunderlin 1999; Pagiola 1999). Because of its inability to get the government to change its forest policies, the World Bank withdrew from this sector in the late 1980s, and it only returned after the New Order government fell.

[2]After nearly five years, a series of World Bank projects with the Indonesian environmental management agency, BAPEDAL, ended with limited success (interview with World Bank staff in BAPEDAL, 1996).

[3]For a discussion of the government's distrust of markets and foreign investment, see Gillis (1984), Timmer (1975), and Hill (1996b). For a discussion of the continuing pursuit of import substitution despite some change in the policy environment toward export promotion, see Rock (1999). A senior official of the State Ministry for Population and the Environment stated that the government has established a team in the Coordinating Ministry for Production and Distribution to prepare Indonesian manufacturers for what is coming in the way of international competition in the twenty-first century. He said the big fear is that because Indonesian industry is so inward looking, most are unaware of international trends (interviews with an official of the State Ministry for Population and the Environment in Jakarta, 1996).

[4]For a discussion of the New Order government's authoritarian politics, see Mackie and MacIntyre (1994). Because of this, the minister of the State Ministry for Population and the Environment supported the development of environmental NGOs, environmental education, and environmental awareness. He also stressed the importance of public participation in environmental impact assessments (EIAs) and the ability of the public to sue polluters (Cribb 1990, 1131).

[5]For a discussion of patrimonial networks linking big business and government and military officials, see MacIntyre (1994), Robison (1986), and Rock (1999).

[6]The political turmoil in Indonesia—the fall of the Soeharto government and ineffectiveness of the Habibie and Wahid governments—means that PROPER PROKASIH has fallen on hard times. Data on plants in the program are still collected, but they have not been acted on or publicly reported since 1997 (communication with Virza Sasmitawidjaja, July 16, 2001).

[7]With the onset of the economic crisis and collapse of the Soeharto government, both the PROPER program and the small-scale monitoring and enforcement program in Semarang collapsed. Neither has been restarted.

[8]This is based on a discussion in 1999 with Shakeb Afsah, the designer of PROPER PROKASIH's computer-based environmental rating system.

[9]This was true for both PROKASIH and PROPER PROKASIH. In fact, the president was kept informed of the ratings in PROPER PROKASIH before those ratings were made public (interview with senior official of BAPEDAL, 1996).

[10]For a discussion of the adoption of one of these programs in Colombia, Papua New Guinea, and the Philippines, see World Bank (2000a, 73–75).

[11]The small-scale monitoring and enforcement program launched by the mayor of Semarang (Aden and Rock 1999) is a notable exception.

[12]State ministries are unique entities in Indonesia. They do not have line authority or implementation responsibilities. They have limited budgets and small professional staffs. Their role is to advise, cajole, and coordinate. The MOI's directive entitled "Prevention and Handling of Environmental Pollution from Industry" established, at least in principle, an EIA requirement for new industrial activities (O'Connor 1994, 7).

[13]Yet the act did not create the legal authority for the environmental ministry to set standards and to monitor and enforce them.

[14]At this point in time, the SMPE had no real alternative to relying on EIAs. As a state ministry, the SMPE had no implementation or line authority, and it lacked the budget and personnel to do more than to advocate, lobby, and cajole. It is interesting that the World Bank commended Indonesia for pioneering this attempt to internalize environmental considerations into the activities of line agencies. As a World Bank report on Indonesia stated, "If Indonesia is successful in this effort to broaden and decentralize environmental responsibilities and concerns, this will be a significant achievement and its example may serve as a model for other countries" (World Bank 1989, 126).

[15]These were covered by Regulation 29, which stated that EIAs would be required for all new and existing projects thought to have a significant environmental impact. The regulation specified that EIAs would cover social, cultural, and environmental effects. It identified five separate documents for each EIA—from one that defined the terms of reference for an EIA, to one that offered a preliminary assessment of environmental effects, to one that spelled out plans for monitoring and reporting on a project's environmental impact. Regulation 29 strongly suggested that EIAs would need to be closely linked with preproject feasibility studies, project designs, and technical and economic analyses of projects. The regulation specified review of EIAs within 90 days of submission of the EIA report by a multiagency AMDAL (Environmental Impact Assessment) Commission (World Bank 1994a, 184–185).

[16]Fifty percent of the BOD load from PROKASIH plants was accounted for by 10% of those plants; 20% of these plants accounted for 75% of the BOD load (Afsah et al. 1995, 18).

[17]This appears to be particularly true of the larger plants responsible for much of the industrial BOD load. Twenty large plants were responsible for the bulk of the reduction in the aggregate BOD load of PROKASIH plants between 1990 and 1994 (Afsah et al. 1995, 21).

[18]The minister did this by announcing the names of those plants not in compliance with their pollution reduction agreements at a speech made to the Jakarta Chamber of Commerce and the press (O'Connor 1994, 135).

[19]Reputational regulation refers to a type of regulation whereby disclosure of a plant's exemplary or poor environmental performance redounds to the plant's benefit or loss. For a discussion of reputational regulation in the United States, see Arora and Cason (1995), who analyze the U.S. Environmental Protection Agency's (U.S. EPA's) "33/50" program. The "33/50" program was a voluntary, but not legally binding, pol-

lution reduction program aimed at 17 toxic chemicals. The aim of the program was to get participating plants to agree to reduce their use of these chemicals by 33% by 1992 and 50% by 1995. Arora and Cason suggest that these goals were achieved.

[20]This program granted local police the authority to investigate citizen "complaints" by taking samples of air and water emissions (Afsah and Vincent 2000, 161).

[21]In 1988, a branch of Indonesia's Legal Aid Society in the city of Medan on the island of Sumatra sued the government for damages to local residents from dust pollution associated with road construction and maintenance. Other NGOs sued several private firms for river pollution associated with wood-processing activities (Cribb 1990, 1133).

[22]The incident involved pollution from an upstream industrial estate that destroyed rice paddies and shrimp ponds in a downstream village along the Tapak River. The Legal Aid Society filed suit on behalf of the affected villagers, and WALHI organized a consumer boycott against the polluting firms. BAPEDAL was brought into the fracas because some of the polluting firms were participants in PROKASIH. In the end, a negotiated settlement was worked out that compensated villagers for their losses, created a fund for village development projects, and led the offending firms to install pollution control equipment to treat their wastewater (World Bank 1994a, 202–203).

[23]Before the recent financial crisis, wastewater emissions charges in industrial estates averaged 3,000 rupiah per square meter (Ruzecki 1997, 18).

[24]When BAPEDAL was created, it inherited PROKASIH from the SMPE.

[25]At the time PROPER PROKASIH was being designed, results from the U.S. EPA's "33/50" program were not available. For a discussion of the results of "33/50," see Arora and Cason (1995).

[26]For example, compliance status was arrived at through the following chain of questions: Do self-monitoring data show the plant in compliance? If the answer is yes, do independently obtained and analyzed samples confirm this? If the answer is yes, the plant is rated as being compliant or blue. If the answer to the first question is yes and the answer to the second question is no, then BAPEDAL authorizes an inspection. Following the inspection, a color-coded rating is applied (Afsah and Vincent 2000, 167).

[27]In early 1995, BAPEDAL sent questionnaires to 350 PROKASIH plants in the 14 industries covered by the national regulations. Of these plants, 176 had sufficient data to be rated. In addition, BAPEDAL invited a number of other facilities to be rated. Eleven agreed, bringing the total to 187 plants (Afsah and Vincent 2000, 167–168).

[28]Community leaders complained about the green rating for one plant; and after investigation, BAPEDAL changed this plant's rating to black (Afsah and Vincent 2000, 169).

[29]For a concise summary of Indonesia's supposedly bifurcated and rent-seeking state, see Rock (1999, 691–694). The ineffectiveness of microeconomic policies has been attributed to the macroeconomic technocrats' lack of influence on and lack of interest in microeconomic policy and to the control of microeconomic agencies, such as the Ministry of Industry and Trade, by those engaged in rent-seeking (MacIntyre 1994, 155; Liddle 1991, 418). For a discussion of the ineffectiveness of microeconomic industrial policies, see Hill (1996b).

[30]Because this meant that consensus within government regarding the environment had not been reached, the president was apparently not willing to sanction the development of a more effective and comprehensive command-and-control environmental agency. This meant that most donor-funded institution- and capacity-building projects in BAPEDAL met with extremely limited success.

[31]Liddle (1991, 415–416) presents compelling evidence that President Soeharto thought in these terms. Because of this, he was often both willing and able to transcend narrow patron–client interests to support important policy changes. In this instance, he may well have done so by sanctioning the use of public disclosure of the environmental performance of major water polluters. He thus signaled to business interests that they had better start addressing the pollution associated with their manufacturing plants. At the same time, because consensus had not been reached within the government regarding the environment, the president was apparently not willing to sanction the development of a more effective and comprehensive command-and-control environmental agency.

[32]For a discussion of each of these crises and the pragmatic responses to them from the New Order government, see Bresnan (1993, Chapters 3, 7, and 10).

[33]But it should be noted that Bulog has also been associated with mismanagement and corruption, and the strategy of the 1970s has bequeathed a rather rigid food pricing policy (Jones 1995; Timmer 1996, 51, 69).

[34]Between 1968 and 1974, farmers received between 19% and 42% of all central bank credit subsidies (MacIntyre 1995, 150).

[35]During the first oil price shock, agriculture received 13% of all development spending. Between 4% and 30% of this was used for fertilizer subsidies (Gelb and associates 1988, 208).

[36]The debate over whether industrial policies in New Order Indonesia were dominated by rent-seeking or successful selective interventions can be found in Hill (1996b), Basri and Hill (1996), MacIntyre (1994), Pack (1994), and Rock (1999).

5

Improving the Environmental Performance of China's Cities

with Fei Yu and Chonghua Zhang

The government of the People's Republic of China is extremely sensitive to international pressure, and it often reacts with outrage to what it considers unwarranted foreign attempts to interfere in its domestic affairs (Okabe 1998, 176).[1] Therefore, it is not surprising that there is no evidence of Chinese pollution management policies being affected by either international economic or political pressure.[2] Instead, the Chinese government's pollution management programs have largely been influenced by internal developments, particularly the partial liberalization of its economy that started in 1979 and the decentralization of decisionmaking that accompanied it.

China's partial economic liberalization and decentralization of decisionmaking sparked a massive rural to urban migration, an enormous increase in economic and industrial output, a huge increase in the consumption of energy, and a corresponding large increase in urban industrial pollution.[3] These events, of course, are well known. Liberalization and decentralization also made local government officials more subject to pressure from citizens within their jurisdictions.[4] Once local citizens and communities began to complain about deteriorating urban environments,[5] the government became aware of the economic and human health costs of environmental degradation, and it set about developing an environmental bureaucracy to clean up the environment.[6]

The way in which the Chinese government established this environmental bureaucracy was, however, determined by China's unique state and domestic politics, particularly its bureaucratic politics. In authoritarian China, bureaucratic politics dominate domestic politics and define the single most important dimension of policymaking.[7] China's public-sector pollution management programs have also evolved incrementally, and they have been most successful when targeted at particularly pressing environmental problems.[8] This combination reflects a pragmatic, trial-and-error approach to problem solving, including environmental problem solving.[9] This is similar to what happened in Indonesia and, to a lesser extent, Malaysia. Moreover, given the vastness of China—a country of more than 1 billion people with 31 provinces and more than 600 cities, 2,000 counties, 100,000 townships, and 1,000,000 villages—it should not be surprising that its pollution management successes have been highly variable (Lieberthal n.d., 3; Jahiel 1998, 759).[10]

It also should not be surprising that the success or failure of pollution management is largely a local, rather than national, phenomenon. Because of this, the focus of this chapter is on one of China's local (city-level) trial-and-error environmental programs: the Urban Environmental Quantitative Examination System (UEQES). This unique program, which annually rates, ranks, and publicly discloses the environmental performance of the country's major cities, was developed and is administered by the national State Environmental Protection Administration (SEPA). But it is implemented locally by city-level environmental protection bureaus (EPBs), in conjunction with mayors and city-level environmental protection commissions (EPCs). UEQES is examined here because it has not received much scholarly attention, it is a quintessential example of a local environmental program, and (as will be demonstrated) its implementation is consistent with a bargaining model of policy implementation that best characterizes the bureaucratic politics of policymaking in China. Before turning to this examination of UEQES, however, we first focus on the evolution of official concern for the environment within the Chinese government.

Official Concern for the Environment

Concern for pollution in China dates from the government's preparation for the 1972 United Nations Conference on the Human Environment in Stockholm. (The rest of this paragraph, unless noted, is drawn from Ross 1988, 137–139.) Although China's position at this conference was confrontational and disruptive, a conference on air emissions control that followed in Shanghai ultimately led to a pilot emissions control project. In 1973, the State Planning Commission held a conference on the environment that identified a need to incorporate environmental considerations into planning. In 1974, the government established a National Environmental Protection Office (NEPO), and similar provincial offices also were established. In 1978, the government amended the constitution to include protection of the environment as one of the most basic commitments made to Chinese society (Panayotou 1999, 434).

Subsequently, four main principles—the polluter pays, prevention first, stronger environmental management, and local control—guided the development of pollution control laws and policies in China (Panayotou 1998, 433; Sinkule and Ortolano 1995, 27). The adoption of the polluter-pays principle signaled to polluters that they, rather than the government, were responsible for control. The focus on prevention first reflected a hope that enterprises could tap large win–win opportunities by preventing pollution. This was seen as appropriate for a poor country such as China that could not afford large expenditures on control. The need to strengthen environmental management expressed a recognition that treating waste was expensive and that enterprises might not treat it unless regulations were enforced. The focus on local control was consistent with the decentralizing tendencies of the post-1979 economic liberalization.

While the government was liberalizing the economy, it also enacted the country's first environmental law, the Environmental Protection Law of 1979 (EPL 1979). EPL 1979 required environmental impact assessments (EIAs) for all new projects, restricted the siting of industrial facilities, established emissions and effluent standards, and introduced pollution discharge fees and the "three simultaneous" program (Panayotou 1998, 434; Sinkule and Ortolano 1995, 27). These discharge fees were applied to wastewater emissions that exceeded emissions standards. The "three simultaneous" program specified that the design, construction, and operation of pollution control facilities must be coordinated with those of new facilities (Sinkule and Ortolano 1995, 32–33).

In 1989, a second environmental law was enacted. The Environmental Protection Law of 1989 (EPL 1989) established an "environmental responsibility system" and introduced a discharge permit system and fines for failing to meet environmental targets.[11] It also inaugurated UEQES, and it mandated limited-time treatment for pollution control and centralized pollution control (Panayotou 1998, 434; Sinkule and Ortolano 1995, 25).[12]

The pollution management instruments created by EPL 1979 and EPL 1989 form the backbone of the government's pollution management program. They have been complemented by the provisions of several other laws, such as the Water Pollution and Control Law of 1984 and the Air Pollution Prevention and Control Law of 1987. Corresponding to these laws are implementation regulations, such as the Implementing Regulations for the Water Prevention and Control Law of 1989 (World Bank 1992a, vol. II, 1–2).

These new environmental laws, regulations, policy instruments, and implementation regulations were developed in tandem with a step-by-step process that enhanced the bureaucratic status and technical capabilities of national, provincial, city, county, and township environmental agencies.[13] Both efforts were important. Without them, there was no prospect that China's public-sector environmental agencies could make an environmental difference. Because of China's unique bureaucratic structure—one that ranks every public organization in the country and limits the influence of an organization to its bureaucratic rank—it was necessary to increase the rank of environmental agencies so that they would have the clout to get others to respond to their directives.[14] Before EPL 1979, environmental agencies were "offices" with low bureaucratic ranks that prohibited them from calling meetings, issuing orders, or gaining access to leading government officials (Jahiel 1998, 767).

Following the passage of EPL 1979, however, several provinces and cities increased the bureaucratic rank of their environmental protection offices (EPOs) by elevating them to EPBs (Jahiel 1998, 767–774). This enabled them to call meetings, issue orders, and gain access to leading local government officials such as mayors. Through 1985, local industrial bureaus (IBs) and factories continued to ignore the stipulations of local EPBs, and NEPO lacked the ability (because of its bureaucratic rank) to mandate changes in the environmental behavior of the IBs that controlled most of China's factories. To facilitate the environmental work of NEPO, in 1984 the State Council created an EPC to increase communication between NEPO and industrial ministries, and it elevated the bureaucratic rank of NEPO to that of a bureau. The latter enabled the National Environmental Protection Bureau (NEPB) to issue orders to provincial EPBs and conduct its own meetings. It also became eligible for independent funding from the Ministry of Finance, and it gained the right to establish an office of international affairs. At the same time, the personnel of the NEPB increased from 60 to 120. These changes led cities to establish their own EPCs and to strengthen their own EPBs.

In 1988, the NEPB was granted independence from the Ministry of Urban and Rural Construction; it was named an agency, the National Environmental Protection Agency (NEPA); its bureaucratic rank was increased; and its staff was increased from 120 to 321. This bureaucratic change gave NEPA the ability to report directly to the State Council. By 1994, NEPA had 10 divisions of second-tier rank with 43 subdivisions, and it was providing local EPBs and EPOs with training and technical advice on pollution control and policy. Many cities followed suit by strengthening the bureaucratic rank and technical capabilities of their environmental agencies, so that by the early 1990s industries were referring to the EPBs as one of the five "hegemonic powers" because of their capability to veto new industrial projects. In 1998, NEPA was elevated to ministerial status and renamed SEPA. Along with this, the State Council Committee on Environmental Protection was abolished, enabling SEPA to turn its proposals into legislation faster.

While the bureaucratic rankings and staffs of national, provincial, and city-level environmental agencies were being increased, the government also invested heavily in strengthening the technical capabilities of these agencies. By 1992, 71 universities or institutions of higher learning were turning out 8,000 graduates a year with environmental specialties (World Bank 1992a, 7). In addition, China accepted a wide range of technical assistance projects to enhance the country's environmental management capacity (NEPA 1996, 174–177). Finally, many of the country's environmental protection agencies drew on a loose affiliation of research institutes, universities, and companies that performed EIAs and designed and sold pollution control equipment (Jahiel 1998, 763).

By 1995, it was clear that China had established a framework of environmental laws and regulations. It had assembled highly dedicated staffs of 88,000 environmental workers in a bureaucracy extending from Beijing to the provinces, cities, counties, and townships; and it possessed a growing cadre of trained environmental specialists to draw on (Lieberthal n.d., 1; Vermeer 1998, 955). As Jahiel (1998, 776) stated:

> Clearly, the past 15 years ... has seen the assembly of an extensive institutional system (for protecting the environment) nation-wide and the increase of its rank. With these gains has come a commensurate increase in EPB authority—particularly in cities.

But given the vastness of China, it should not be surprising that there is wide variability in the capabilities and performance of local, particularly city-level, EPBs. To quote Jahiel (1998, 759) again:

> There is great variability in size, funding, staffing, and even work methods of environmental protection agencies in different parts of the country.... Environmental protection agencies in the wealthier coastal provinces and in large cities tend, as a rule, to have more personnel, be better funded, and be staffed with more technically-trained people than those agencies in the poorer interior parts of the country, smaller cities, counties, and townships.

In addition to the wide variability in the capabilities and performance of local environmental agencies, environmental professionals in China's bureaucracy faced and continue to face two problems. First, the public financing of China's myriad environmental agencies has failed to keep pace with growing personnel and responsibilities. This has left these agencies shorthanded, contributed to an erosion of staff compensation, and forced many agencies to look for their own sources of financing. Several analysts have argued that these developments have hampered agencies' effectiveness by making it difficult for them to attract qualified personnel and forcing them to become self-supporting in ways that undermine their regulatory functions.[15] Second, environmental professionals at all levels of government have found themselves engaged in an intense game of bureaucratic politics that requires hard bargaining with others in the bureaucracy over how much environmental improvement should be allowed.

Bargaining over Environmental Improvement

Policy implementation in China is dominated by vertical and horizontal bargaining within the country's multilayered public-sector bureaucracy (as Lieberthal 1992, Lampton 1992, and Walder 1992 all have observed). "Vertical" bargaining refers to that between two or more organizational units of unequal bureaucratic rank, whereas "horizontal" bargaining refers to that between two or more organizational units of equal rank. Vertical bargaining can occur between different territorial units of government (e.g., between the central government and provinces or between provinces and cities) or within the same territorial unit (e.g., between two units of unequal rank within a city). Negotiations between city governments and provincial officials over retention of a larger share of taxes collected by cities is an example of the first. Negotiations between an enterprise and the IB that has authority over it concerning tax treatment is an example of the second. Horizontal bargaining in cities includes, for example, negotiations among a finance bureau (FB), a tax bureau (TB), and a local bank over the funding of investment projects; and bargaining between an EPB and an urban construction bureau (UCB) over the dust and noise associated with urban construction.[16]

Why does so much bargaining occur over policy implementation? There are three answers to this question. First, the official ranking of positions, organizational units, and functional bureaucracies in China's public sector creates situations in which a number of position holders, organizations, and bureaucracies must reach agreement before policies can be implemented. Agreement might be needed, for example, because provincial governors, who have the same rank as ministers of functional bureaucracies, can block new policy initiatives by central ministries. Or agreement might be needed because the heads of several central ministries, who have the same rank, can block implementation of new initiatives proposed by any one of these ministries. Within cities, agreement might be blocked if IBs of higher rank than local EPBs refuse to take the pollution mitigation actions suggested by the latter.

Second, bargaining has proliferated since the post-1979 decentralization of decisionmaking that strengthened the power and authority of local officials while weakening that of provincial and central government officials.[17]

Lieberthal (n.d., 4–5) argues that this reflects an implicit agreement between different levels of government, whereby each level grants to the level just below it the flexibility needed to grow the economy, but not so much flexibility that the level above loses ultimate control. Because of this implicit deal, he contends that China's political system has become highly negotiated at every level, as key officials spend inordinate amounts of time negotiating for more flexibility while trying to keep higher levels from becoming too restrictive.

Third, bargaining within cities over how to implement particular policies is a logical outcome of the partial, incomplete liberalization of the economy (Walder 1992). On the one hand, this unfinished liberalization has fostered the growth of many industrial enterprises that are owned and managed by city governments. Mayors and other city managers, particularly those on planning commissions (PCs), economic commissions (ECs), FBs, and IBs, are expected to adopt policies that promote the growth, development, and profitability of these enterprises. On the other hand, their ability to promote the growth and profitability of any enterprise is affected by the partial liberalization of input prices. In some instances, state-controlled input prices undermine enterprise profitability; in others, they artificially inflate profitability. City governments have reacted to this by developing simple tax and subsidy rules for individual enterprises.[18] This is important because city revenues—which are used to fund city government, social services, public works, urban infrastructure, environmental regulation, and new investment projects, including pollution control investments in city-owned industrial enterprises—are overwhelmingly derived from taxes on those same enterprises.

Because of the need to balance enterprise growth and development objectives with the other pressing needs within cities where enterprise profits may not reflect objective conditions, decisions to tax an enterprise and fund its investments—including its pollution control investments—depend on intense bargaining among all those involved. Individual enterprises bargain with IBs over the tax treatment and funding of investment proposals. IBs, the FB, the TB, the UCB, the EPB, and banks that make loans to enterprises for new investments bargain over these same issues, as well as over the need to increase investments in urban infrastructure and to invest in pollution control. The mayor's office, the FB, the PC that oversees long-run growth in a city, and the EC that is responsible for ironing out bottlenecks facing enterprises and industries bargain over this same set of issues. As a consequence, the decisions to invest in an enterprise, tax it, and subsidize it are all made collectively and are inextricably intertwined. This preference for collective decisionmaking, and hence for bargained solutions to policy implementation, also reflects a bureaucratic culture that operates on a principle of fairness. Fairness is said to exist when consultation takes place and when the outcome of consultation does not leave any locality or organization without the resources it needs to subsist and to accomplish its duties (Lampton 1992, 39).[19]

Implementation of Environmental Policy

How have the bureaucratic politics of bargaining within China's public-sector bureaucracy affected the effectiveness of national, provincial, and city-level environmental policies designed to curb pollution? Lieberthal (n.d.), Jahiel (1998), and Vermeer (1998) contend that the powerful local actors (mayors, ECs, PCs, and IBs) that control enterprises and environmental regulators still regard environmental outcomes as secondary to economic growth. Because of this, they conclude that it is not surprising that local urban ambient environments are not improving or that EPBs do not have much influence. Although there is some truth to this observation, examination of one of China's most distinctive environmental policies—SEPA's annual rating, ranking, and public disclosure of the environmental performance of the country's major cities—leads to the conclusion that such a view is both too simple and, in important ways, wrong.

After 1989, NEPA has conducted annual quantitative assessments of the environmental performance of the country's major cities. (In 1998, NEPA was elevated to ministerial rank and renamed SEPA; Jahiel 1998, 757. Subsequent references will be to SEPA.) Provincial EPBs have followed suit by extending SEPA's UEQES to a large number of provincial cities.[20] In both cases, quantitative assessments are based on a city's composite score on more than 20 environmental indicators (a list of the indicators is given in Table 5-1 below). Some of the indicators focus on ambient environmental quality, some focus on the level of development in urban environmental infrastructure, and yet others focus on indicators of urban environmental management (pollution control). Scores for each indicator are weighted and summed to yield a composite score. Cities are ranked on the basis of their composite scores, and SEPA publishes ranks and scores in its annual environmental yearbook (NEPA 1996, 129). Some provinces and cities also publicize scores and rankings in newspapers and on radio and television.[21]

Unfortunately, relatively little is known about UEQES' effects on either city-level environmental management or urban ambient environmental quality. Yet there are four reasons to suspect that it might be making a difference. First, there is substantial evidence that some indicators of urban ambient environmental quality have either improved or at least not deteriorated in the country's major cities, despite significant growth in population and economic activity. A recent study of air quality in China's largest cities found that concentrations of total suspended particulates (TSP) in the air declined from roughly 800 micrograms per cubic meter ($\mu g/m^3$) to 500 $\mu g/m^3$ in north

China and from 600 μg/m^3 to 400 μg/m^3 in south China. This was achieved by cutting the energy intensity of the economy by nearly 50% and by reducing the TSP intensity of gross domestic product (GDP) from 34.3 tons per million yuan of GDP in 1981 to 9.8 tons per million yuan of GDP in 1993 (Wang and Liu 1999, 381, 385).

Second, most studies of China's other major environmental policy initiatives—such as EIAs, the "three simultaneous" program, the pollution levy, and the pollution discharge system—have concluded that they have had relatively little effect on the environmental behavior of polluters.[22] This leaves open the possibility that UEQES may be working. Third, the administration of UEQES and the system of environmental target responsibility to which it is linked are more consistent with an integrated approach to urban environmental management and with the bargaining model that characterizes environmental policy implementation in China than are any of the country's other environmental policy instruments.[23] Fourth, evidence from elsewhere, including Indonesia, another low-income economy, suggests that publicly disclosing environmental performance can have a significant effect on behavior (Afsah and Vincent 2000). This combination increases the likelihood that UEQES may be making a difference.

Because of this possibility, the remainder of this chapter describes the UEQES index, examines how it is used for environmental management by focusing on its application in five cities (Beijing, Changzhou, Nanjing, Shanghai, and Tianjin), and evaluates the effect of UEQES on ambient environmental quality in the five cities.

The UEQES Index

Before we begin to describe the index used in the Urban Environmental Quantitative Examination System, it is important to note that at least three urban environmental indices are used in China. One index is used by SEPA for quantitative assessment in UEQES. A second index is used by SEPA in its sustainable cities program.[24] A third index—or, more precisely, a third set of indices—is used by provinces to rate and rank provincial cities not evaluated by SEPA. These provincial indices, which vary from province to province, are modified versions of SEPA's UEQES index;[25] these also are living indices, in the sense that SEPA and provincial EPBs constantly modify them. Modifications include adding and dropping indicators, changing the formula used to calculate the score for an indicator, and changing the weight of an indicator or a group of indicators in the overall index. How this happens and what it means are best demonstrated by giving an example.

Table 5-1 lists the indicators used in the UEQES index from 1991 to 1995. Note that the first 6 indicators focus on ambient environmental qual-

Table 5-1. Indicators in the Index for China's Urban Environmental Quality Examination System, 1991–1995

Indicators
Indicators of Ambient Environmental Quality
Annual average total suspended particulate concentration (micrograms per cubic meter, or μg/m³)
Annual average sulfur dioxide concentration (μg/m³)
Drinking water source compliance rate (percent)
Ambient surface water chemical oxygen demand concentration (micrograms per liter, or μg/l)
Ambient noise level in urban areas (decibels)
Ambient noise level at major roads in urban areas (decibels)
Indicators of Pollution Control
Percentage of urban area enforcing smoke and dust control
Household use of briquettes (percent)
Compliance of industrial air emissions (percent)
Compliance of automobile air emissions (percent)
Wastewater discharge per 10,000 yuan of output
Industrial wastewater treatment rate (percent)
Percentage of treated industrial wastewater that meets emissions standards
Rate of recycling of solid industrial waste
Rate of industrial solid waste treated and disposed
Percentage of urban area under ambient noise control
Indicators of Urban Infrastructure
Rate of treatment of urban garbage (percent)
Rate of household use of gas (percent)
Rate of household use of centralized heating (percent)
Percentage of urban area devoted to green space
Rate of municipal wastewater treated (percent)

Source: Communication from State Environmental Protection Administration, 1999.

ity, the next 10 focus on pollution control, and the last 5 focus on urban environmental infrastructure. The indicators of ambient environmental quality account for roughly 30% of the total weight in the index, the indicators of urban environmental management and pollution control account for 50%, and the indicators of urban environmental infrastructure account for the remaining 20%. For each indicator listed in Table 5-1, scores were determined using a formula, a weight, and high or low values. For example, the current formula for the TSP ambient air quality indicator is given by TSP = $4^{*}(0.6 - X)/0.42$ for cities in north China and TSP = $4^{*}(0.5 - X)/0.42$ for cities in south China. X is the actual annual daily mean concentration of TSP in micrograms per cubic meter of air, such as 0.700 or 700 μg/m³. The number 4 is the weight of TSP in the overall UEQES.[26] The current maximum and minimum values for TSP for cities in north China, which rely heavily on coal, are 0.6 $\mu g/m^3$ and 0.18 $\mu g/m^3$. For cities in south China, which are less dependent on coal, current maximum and minimum values are 0.5 $\mu g/m^3$ and 0.08 $\mu g/m^3$.

Although the use of formulas with weights and maximum or minimum values for individual indicators may appear confusing, it is in fact quite ingenious. This is best illustrated by examining the TSP indicator more closely. If a city in north China has an average annual daily mean TSP concentration of 600 $\mu g/m^3$, its score on the TSP indicator will be zero.[27] Conversely, if its average annual daily mean concentration of TSP is 180 $\mu g/m^3$ or less, its score will be 4.[28] Because TSP concentrations in the large cities of north China have fallen from roughly 800 to 500 $\mu g/m^3$, this formula takes current ambient environmental reality into account by punishing cities with TSPs above 600 $\mu g/m^3$ and rewarding those with TSPs lower than that.[29]

Because SEPA treats UEQES as a living index, it can and does modify individual indicators as reality changes. In the case of the TSP indicator, at least two changes have occurred. First, SEPA has increased the weight attached to ambient indicators. Second, it has "tightened standards" in this index by changing the maximum and minimum values of the TSP indicator.[30] Because of these changes, Beijing's score on ambient indicators and its overall score and ranking have been falling as its ambient environmental quality has improved at a slower rate than that of other large cities in north China. It is interesting that the mayor of Beijing may have reacted to this decline by agreeing to take a number of significant steps to improve ambient air quality.[31] How and why this is happening in Beijing and in other cities will be examined later in this chapter.

The use of simple formulas, weights, and maximum or minimum values applies equally as well to the other indicators in the UEQES index. Changes in each also occur as SEPA ratchets up expectations over time. For example, the current formula for the sewage treatment rate (STR), one of the indicators of urban environmental infrastructure, is $STR = 4(X/0.4)$, where 4 is the weight, X is the current percentage of sewage treated, and 0.4 or 40% is the maximum value for the percentage of sewage treated.[32] A city thus can get a maximum score on this indicator of 4 if 40% of its sewage is treated. It gets a zero score if none of its sewage is treated. As with the TSP indicator, SEPA has been "tightening this standard" by increasing the value of X over time.

Because (as was noted) SEPA treats UEQES as a living index, it has made other changes in indicators. This happened recently with the indicator measuring the compliance rate of automobiles with emissions standards (CRAES). The CRAES indicator used to be calculated by the formula $CRAES = 3*(X - 30)/50$, where the number 3 was the weight, X equaled the compliance rate with emission standards as measured in vehicle emissions tests in

Table 5-2. Scores on the Index of 37 Chinese Cities for the Urban Environmental Quality Examination System, 1995

City	*Score*	*Rank*	*City*	*Score*	*Rank*
Tianjin	84.37	1	Harbin	77.01	20
Beijing	83.62	2	Taiyun	76.04	21
Suzhou	83.25	3	Jinan	75.18	22
Haikou	82.2	4	Fuzhou	74.53	23
Dalian	82.08	5	Shenyang	73.80	24
Shijiazhuang	82.05	6	Hefei	72.91	25
Guangzhou	81.81	7	Nanning	71.12	26
Shenzhen	81.63	8	Changchun	70.88	27
Chengdu	81.37	9	Nanchang	70.71	28
Shanghai	80.99	10	Ningbo	70.17	29
Hangzhou	80.58	11	Kunming	69.37	30
Wuhan	79.52	12	Guiyang	68.78	31
Nanjing	79.50	13	Yinchuan	67.28	32
Xi'an	79.29	14	Xining	66.51	33
Zhengzhou	79.21	15	Lanzhou	65.29	34
Guilin	78.83	16	Chongqing	64.39	35
Changsha	78.34	17	Urumqi	60.59	36
Qingdao	78.23	18	Hohhot	56.91	37
Xiamen	78.16	19			

Note: Because the sum of the weights on all the indicators adds up to 100, and because the formulas for the indicators are designed to yield a maximum score equal to the indicators' weights when a city achieves the maximum or minimum value for an indicator, the scores on the index range from 0 to 100. For example, given the formula for the total suspended particulates indicator for cities in north China ($TSP = 4*(0.6 - X)/0.42$), a city can achieve a maximum score of 4 (the weight on this indicator) on this indicator if annual average daily TSP is 180 micrograms per cubic meter or less.

Source: NEPA 1996, 167.

shop tests, and the number 30 was the minimum value for the percentage of automobiles tested that met emissions standards.[33] Once SEPA discovered "corruption" in shop test results, however, it modified the way X was calculated. Instead of simply relying on reported results from shop tests, it used the formula $X = 0.5*(X_1 + X_2)$, where X_1 was the compliance rate from shop tests and X_2 was the compliance rate from street-level spot checks of automobile emissions.

Table 5-2 lists the scores and ranks of the 37 cities evaluated by SEPA under the UEQES index system in 1995, which ranged from a low of 56.91 for Hohhot to a high of 84.37 for Tianjin. Although there are no statistical studies of the relationship between nonenvironmental characteristics of cities and their scores and ranks, both city-level EPBs and SEPA suspect that such relationships may exist. A cursory examination of the scores and rankings of cities for 1995 suggests that they may be right. Note, for example, that old, poor inland cities such as Kunming, Xining, and Chongqing score and rank lower than young, rich coastal cities such as Suzhou, Guangzhou, and Shenzen.[34] This has led some to ask whether the index treats all cities fairly. The issue of fairness in the index will be discussed in more detail in the next section.

The UEQES Process

At least three different questions can be asked about the administration of the examination process embodied in UEQES:

- How does SEPA organize and manage the process?
- What role do city governments play in the process?
- How, if at all, does the process affect local urban environmental management?[35]

Because we are most interested in the role that city governments play in the UEQES process and whether and how the annual examination process affects urban environmental management, we focus on the second and third questions. Our answers to these two questions are based on discussions with senior officials in EPBs in five cities and on the literature on policy implementation in China, including that on urban and environmental policy. We begin by describing how UEQES is implemented. Then we show with a set of examples how local EPBs in the five cities used the UEQES process to attract and hold the attention of mayors and to successfully bargain with vaunted local economic bureaucracies over environmental issues.

The ability of city-level EPBs to attract and hold the attention of mayors and bargain effectively with local economic organizations can partially be explained by the institutional developments described above that enhanced the bureaucratic status of EPBs and increased their access to local government leaders. (The discussion that follows is based on Jahiel 1998.) Three of those institutional developments are particularly noteworthy. As was described above, in 1979 the government enacted its first environmental law. Among other things, EPL 1979 increased the bureaucratic rank of city-level EPOs and renamed them EPBs. This made them first-tier organizations within China's bureaucratic ranking system, enabling them to call meetings; to issue orders; and, for the first time, to have direct access to local government leaders such as mayors.

Because local IBs and factories continued to ignore national, provincial, and local EPB requests to mitigate pollution, the State Council, at the second National Environmental Protection Conference in 1983–1984, created an EPC to facilitate direct communication between the national EPB and industrial ministries. This action was followed later in 1984 by the State Council's decision to increase the bureaucratic rank of the national EPB and to change its name to the NEPB. These changes enabled the NEPB to conduct its own meetings and to issue orders to provinces and cities. Among these orders were recommendations to cities that they establish local EPCs and strengthen EPBs. Many did both. By 1988, when the NEPB became the independent NEPA—with an even higher bureaucratic rank—many cities had achieved sufficient bureaucratic authority and technical capacity, along with direct access to the local economic bureaucracy through the local EPC, to bargain with NEPA.

Despite these institutional developments, senior officials in the EPBs in the five cities stated that they continued to have serious difficulty getting taken seriously by industrial enterprises and their representatives in the economic bureaucracy or in the urban construction and service bureaus.[36] As officials of one city-level EPB stated, before the establishment of the annual UEQES process and the environmental target responsibility system to which it is attached, "we were kind of vague. We collected data at the point of lowest resistance, where we could get it. The system was good but the people were not."

All this appears to have changed with the implementation of the UEQES process. Senior officials of every EPB interviewed commented that, first and foremost, the process—particularly the scoring, ranking, and publishing of results by SEPA in its annual yearbook—captured and held the attention of mayors.[37] Suddenly mayors wanted to know why their city scored and ranked lower than other cities. They wanted to know what was in the index and how it worked—that is, how and why it led to a particular score. They also wanted to know what could be done to increase a city's score and rank, and they wanted to know what this would cost. These questions led officials in the EPBs interviewed to study SEPA's indicators and the formulas, weights, and high or low values used to calculate the score for each indicator. In some instances, this was done to determine the most inexpensive way to increase a city's overall score on the UEQES index.[38] Sometimes this led to new research on these topics once SEPA increased the weight of an indicator, "tightened it," or otherwise changed it.[39]

The annual UEQES process also provides city-level EPBs with their first real opportunity to engage with other organizations and units in city government, particularly those in vaunted economic bureaucracies, over environmental issues. How this works is best demonstrated by describing the process used by cities to implement UEQES. To begin with, every major city in China now has a five-year plan and an annual plan to improve its environment. The production and implementation of these plans are under the authority of the city's EPC. Normally, the chairperson of the EPC is either the mayor or a vice mayor. Other members of the EPC include senior representatives from the EC, IB, FB, TB, UCB, urban services bureau (USB), and several others.

The implementation of UEQES within a city is led by the EPC, but the local EPB serves as the key technical office for the EPC. The whole process begins when the mayor asks each line agency and sector or district within the city to put together a report that evaluates their performance relative to last year's environmental targets and proposes environmental targets for the current year. The EPB takes this information; analyzes it; and consults with appropriate line agencies, sectors, and districts regarding last year's performance and this year's targets. The EPB officials who were interviewed stated that this is a very delicate job. It requires EPBs to balance the need to show tangible improvements with the need to be realistic. They do this by increasing their understanding of what it takes to improve the score for a particular indicator and by bargaining and negotiating with line agencies, sectors, and districts over how large improvements in these indicators should be.

Once individual line agency, sector, and district reports with targets for the current year are agreed to by the appropriate line agency, sector, district, and EPB, the EPB rolls this up into a projected overall UEQES target score for the current year. All of this is included as part of a larger background report that the EPB prepares for the EPC early each year. This report describes what the city did to improve its score last year, identifies the major environmental problems in the city, proposes targets by indicator for the current year, and assigns responsibility for meeting those targets to the appropriate line agencies. A comprehensive description of how UEQES operates in a city and a big table showing who is responsible for what in meeting the current year's targets and how UEQES relates to SEPA's other environmental policy instruments are included as part of the report.

After the report has been submitted, the EPC holds its first general meeting of the year on UEQES. At this meeting, the EPC reviews past progress; examines current environmental problems; agrees on this year's target score for the UEQES index; and discusses and assigns responsibility for meeting targets for specific indicators to specific line agencies, sectors, and districts. This meeting culminates in the signing of contracts between each sector of government and the mayor for the environmental target responsibility system. These contracts serve as the basis for the mayor's contract with the provincial governor for the same system. Once all of this has been completed, the EPB publishes a summary report on the EPC meeting in its monthly bulletin. This report includes recommendations that have been given, consensus that has been reached, and decisions that have been made.

The EPB distributes the report to each of its divisions and asks them to prepare a plan to ensure that targets will be met. This divisional review may lead to some redrafting. If this happens, the EPB takes the report back to the EPC for final approval. Once final approval is given, the EPC holds a larger "mobilization" meeting to get all of the sectors and districts involved in implementation.

During the year, the EPB continues to play an important role in implementing the UEQES process. For example, senior representatives of the EPB meet approximately two times a year with each line agency (e.g., the UCB, USB, and various IBs) to evaluate performance relative to the targets in the contract between the sector and mayor. The results of these meetings are rolled up into midyear and end-of-year EPC reviews of progress by line agency, sector, and district. The EPB prepares the background reports for these reviews. It is interesting that the Personnel Department, an office of the Communist Party, participates in these reviews.

One other aspect of the UEQES process deserves mention. As was hinted at earlier in the chapter, the whole process is tightly linked to the environmental target responsibility system.[40] This system is designed to make provincial governors responsible for the environmental quality in the provinces and mayors responsible for it in the cities. It is also designed to get governors and mayors to develop a better balance between growth and the environment. This is done by having governors and mayors sign environmental performance contracts with officials one level higher up in the bureaucratic hierarchy. These contracts specify that particular outcomes will be achieved within specified time periods. The actual parameters included in contracts are initially established by the central government and then distributed to provinces and cities. Numerical targets for each parameter at each level of government are the result of bargaining. Once a governor and mayor reach agreement on the target, it is made public. As with the UEQES process, the results of assessing performance relative to the target are made public. This assessment can and has resulted in rewards or punishments.[41]

The way in which the environmental responsibility contract system is linked to UEQES is best explained with an example. In 1993, the mayor of Nanjing signed a five-year (1993–1998) environmental performance contract with the provincial governor.[42] The contract is modeled on the UEQES index. One part of the contract focuses on ambient environmental quality, one on pollution control, and one on urban environmental infrastructure. With respect to UEQES ambient environmental indicators, the mayor agreed to reduce TSP levels from about 350 $\mu g/m^3$ in 1993 to less than 220–250 $\mu g/m^3$ in 1998. He also agreed to improve the quality of surface water in the Yangtze River flowing through the city so that it meets the Class 2 standard, and he agreed to clean up Xuan Wu Lake so that it meets the Class 5 standard for surface water. With respect to the pollution control indicators in

UEQES, the mayor agreed to reduce industrial wastewater emissions to less than 350 m^3 per 10,000 yuan of industrial output. He also agreed to increase the industrial wastewater treatment rate to 70%, to reduce chemical oxygen demand (COD) levels in surface water by 10%, and to limit dust to no more than 178,000 tons a year. With respect to the indicators of urban environmental infrastructure in UEQES, the mayor agreed to 4 air pollution control projects, 15 water pollution control projects, and 1 solid waste project.

Each of these targets in the mayor's contract with the provincial governor was arrived at through a bottom-up planning process in the city that assigned each indicator in UEQES and in the mayor's contract with the provincial governor to one or more sectors or districts in Nanjing. For example, the urban environmental investment projects in air pollution control were assigned to the EC. The EC worked with its various IBs, and these in turn worked with individual industries and plants under their authority to develop plans to increase investment in air pollution control at specific plants. Because the IBs negotiate with individual plants over future investment plans, including plans for pollution control expenditures, they know which plants are planning to expand, renovate, and make air pollution control investments.

Once agreement is reached on future plant-level investment in air pollution control, the IBs bargain with the FB over tax treatment of proposed investments, with the banks over loans for investments, and with the EPB over the use of pollution levy funds to subsidize part of the air pollution control investment. Once these negotiations are completed, IBs develop estimates of the number of air pollution control projects and the cost of those projects in their sectors for the next five years. The EC adds up each of these to develop an overall estimate of the number of projects and their cost during the next five years. It then signs an environmental target responsibility system contract with the mayor specifying the number and cost of projects it will undertake during the next five years.[43] That contract becomes the basis for the mayor's contract with the provincial governor. Several senior representatives of EPBs in the five cities stated that in this way the environmental target responsibility system provides the technical basis for a city's target score on individual UEQES indicators and for the overall target score.[44]

This process is repeated for each UEQES indicator in each sector and district in the city. For example, if the mayor's environmental contract includes an increase in the percentage of municipal wastewater treated, before he agrees to this, the UCB and FB, among others, will have agreed to include this project in the city's overall investment budget. Once this has been agreed to, the UCB signs a performance contract with the mayor that specifies the percentage increase in wastewater treatment that will be achieved during the next five years. Once this projected outcome has been agreed to, the mayor includes it as part of his contract with the provincial governor.

The Impact of UEQES

What evidence is there that the UEQES process actually affects urban environmental management and ambient environmental quality? There are two answers to this question. First, we offer several concrete examples of how cities have used UEQES to affect integrated urban environmental management. Second, we look more closely at changes in UEQES indicators in two cities over time. Although this evidence is not definitive, it suggests that UEQES is making an environmental difference.

We begin by describing how the Tianjin EPB recently used UEQES to reduce the city's TSP levels.[45] In the 1980s, its TSP levels averaged about 500 $\mu g/m^3$; today, they are 300 $\mu g/m^3$. EPB officials attribute the decline in TSP levels to an increased percentage of households using briquettes and gas for home cooking, of households with centralized heating, and of major boilers in the city that have installed scrubbers. The officials attribute each of these increases to the implementation of UEQES. They describe how this was accomplished: The city has 18 districts or counties, and the mayor and the city's EPC used a bottom-up planning process to set target TSP concentration levels.

This bottom-up planning process took account of how many households were in a district or county, what percentage of households cooked with briquettes and gas, what percentage heated with central heat, how many industrial boilers operated in a district or county, and what percentage of those installed and used scrubbers. Following the completion of this inventory, each district or county developed estimates, in conjunction with the USB, EC, and several IBs, of what it would cost to extend household use of briquettes as well as gas and central heat, and what it would cost to install scrubbers on boilers. After making several assumptions about how much expansion in each area could be financed by the city in any year, the city developed annual and five-year targets for each UEQES parameter. The local EPB used targets for these parameters to set new targets for TSP concentration levels within the city. These were set at no more than 250 $\mu g/m^3$ for counties, 310 $\mu g/m^3$ for the six dirtiest districts, and 250–310 $\mu g/m^3$ for the remaining districts.

Once these new targets were agreed to, the EPB used its real-time monitoring of TSP in each district and county to track performance relative to each target each year. When performance in a district fell below the target, the EPB would work with the district EPB to figure out why. As EPB officials stated, failure to meet a target could often be pinpointed to one plant. The EPB would then negotiate with the IB that had authority for the plant over how it could reduce its emissions so that the city could meet its TSP target. In difficult cases, EPB officials called on the mayor, EPC, or city personnel department for help. The result of this target-setting and -tracking process is that the city's score on the TSP indicator has gradually risen from less than 1 point to almost 3 points.

The Nanjing EPB used UEQES to clean up a large lake (3.6 square kilometers), Xuan Wu Lake, which is located in a central park in the city.[46] In the 1980s, this lake suffered severe eutrophication; it had turned septic, many fish had died, and it emitted a particularly foul odor. The public complained about the lake, and the problem was finally brought to the attention of the mayor and the Nanjing EPC. The mayor and EPC established a special study group to propose recommendations for cleaning up the lake. The group recommended a number of projects. One required intercepting and diverting sewer water going into the lake. A second required industrial polluters who were dumping their water into these sewers to meet standards for sewer water discharge. A third entailed dredging the lake bottom. Others included capturing and diverting rainwater runoff, relocating a zoo that was discharging its wastes into the lake, and changing the ecology of the lake so that submerged grass could take up nitrogen and slow eutrophication.

The study group concluded that if these recommendations were implemented, the water in the lake could meet at least Class 5 surface water quality standards. On the basis of the group's recommendations and agreement within the city to finance and implement them, the mayor included cleanup of the lake to the Class 5 level in his environmental contract with the provincial governor. By the end of 1998, the city had invested approximately 120 million yuan in the lake cleanup project. The EPB's vice director believes that the lake will be fully restored in three years. Because of these investments, the mayor will soon be able to fulfill the part of his contract with the provincial governor that pertains to the lake's ambient environmental quality.

The EPB in Changzhou used UEQES to reduce noise levels along major trunk roads, to increase the municipal wastewater treatment rate, and to improve the compliance rate for drinking water quality.[47] In each instance, actions taken by the city to improve performance resulted in increases in the city's UEQES score. During the eighth plan period (1991–1995), noise levels on major trunk roads averaged 77 decibels (dB), and the city's score on this indicator was zero. The EPB and UCB studied this problem and attributed the high noise levels to the poor quality of the road, of cars, and of traffic management (which also contributed to congestion). The UCB and EPB developed an investment strategy to reduce noise levels that required widening roads at major intersections and building overpasses to reduce congestion. After these changes were implemented, noise levels fell from 77 to 71.2 dB, and the city's score on this indicator rose from 0 to 2 points. Because senior officials in the local EPB consider any increase of 0.01 or more in the overall UEQES index significant, the increase of 2 points was considered a major success.

The city was also successful in increasing its score on the indicator of drinking water quality. After the EPB studied performance on this indicator, it recommended that the city draw less of its drinking water from the Grand Canal and more from the Yangtze River. This required moving the major intake facility for surface water, but this was an expensive endeavor, and the city's USB opposed this move. After financing for this was agreed upon, however, the city moved the facility from the canal to the river. This increased the compliance rate on the intake source of water quality for the drinking water parameter from 75% to 96%, and the city's score on this parameter rose from 2.7 to 4.6 points.[48]

Changzhou was less successful in increasing its score on the indicator of municipal wastewater treatment by increasing its investment in centralized treatment. In the early 1990s, the local EPB proposed increasing investment in municipal wastewater treatment, and during five years treatment capacity increased from 18,000 to 75,000 cubic meters a day. The mayor, EPC, and EPB expected this to result in a 2-point increase in the score on this parameter, but it only resulted in a 1-point increase. The small increase in score on the wastewater treatment indicator disappointed everyone involved and caused some problems for the EPB. For one, the EPB had to explain to the mayor and EPC why the increase was so small. After studying this problem, the EPB concluded that this was the result of SEPA "tightening the standard" on the wastewater treatment indicator. This incident has led the Changzhou EPB to study each indicator more closely and to ask whether UEQES is fair to all cities.

How have these kinds of actions affected the environmental performance of cities over time? The answer to this question can be found by looking more closely at the performance on each indicator in the UEQES index of two large cities—Beijing and Shanghai—between 1988 and 1995. Raw data for each indicator in each city are given in Tables 5-3 through 5-6. Four observations stand out. First, ambient environmental quality is very poor in both cities. TSP levels range from 264 to 358 μg/m^3 in Shanghai and from 348 to 450 μg/m^3 in Beijing. Sulfur dioxide (SO_2) levels range from 53 to 110 μg/m^3 in Shanghai and from 90 to 132 μg/m^3 in Beijing. COD levels in surface water range from 458 to 761 micrograms per liter (μg/l) in Shanghai and from 605 to 705 μg/l in Beijing. Second, except for COD levels in Shanghai and recent improvements in SO_2 levels in Shanghai, there is no noticeable improvement in the ambient indicators. Third, both cities have made major investments in urban environmental infrastructure. For example, municipal

Table 5-3. Integrated Urban Environmental Quality Index for Beijing, 1988–1990

Item examined	*Unit of measure*[a]	*1988*	*1989*	*1990*
Items Related to Environmental Quality				
Total suspended particulates, annual average	μg/m³	450	399	407
Sulfur dioxide, annual average	μg/m³	100	99	122
Drinking water source compliance rate	%	95.9	93.2	94.2
Ambient surface water chemical oxygen demand level	μg/l	630	605	629
Average ambient noise level in the urban area	dB	60	58.6	58.8
Average noise level at major roads in the urban area	dB	72	72.1	71.4
Items Related to Pollution Control				
Urban area enforcing smoke and dust control zone program	%	42	78.1	100
Household use of briquettes	%	55	53	56
Compliance of industrial air emissions	%	60.9	82	94.6
Compliance of automobile air emissions	%	100	85	100
Wastewater discharge quantity per 10,000 yuan of productivity	m³	113	108	111
Industrial wastewater treatment rate	%	30	41.1	45.3
Rate of treated industrial wastewater in compliance with standards	%	78	74.2	73.6
Rate of industrial solid waste used	%	50	48.8	50
Rate of industrial solid waste treated and disposed of	%	19	21.1	15.8
Items Related to Development of Urban Infrastructure				
Rate of household use of gas in city	%	90	85.5	83.9
Rate of households with centralized heating	%	22	19.3	21.1
Rate of municipal wastewater treatment	%	7.4	6.7	7.2
Rate of urban garbage centrally managed	%	83	100	100
Average green space per person	m²	5.8	6	6.14

[a]Units of measure: μg/m³ is micrograms per cubic meter, μg/l is micrograms per liter, dB is decibels, m³ is cubic meters, and m² is square meters.

Source: Urban Environmental Management Division, State Environmental Protection Administration.

Table 5-4. Integrated Urban Environmental Quality Index for Beijing, 1991–1995

Item examined	*Unit of measure*[a]	*1991*	*1992*	*1993*	*1994*	*1995*
Items Related to Environmental Quality						
Total suspended particulates, annual average	μg/m³	353	380	348	395	377
Sulfur dioxide, annual average	μg/m³	126	132	120	110	90
Drinking water source compliance rate	%	97	95	96	98	99
Ambient surface water chemical oxygen demand level	μg/l	705	638	670	620	650
Average ambient noise level in the urban area	dB	59.7	68.5	57.8	56.9	57.1
Average noise level at major roads in the urban area	dB	72	71.6	71.7	71.7	71.7
Items Related to Pollution Control						
Urban area enforcing smoke and dust control zone program	%	100	100	100	100	100
Household use of briquettes	%	68.77	70.91	76.5	82.8	85.2
Compliance of industrial air emissions	%	99.45	99.68	92.3	94.2	96.7
Compliance of automobile air emissions	%	68.55	78.45	79.4	80.5	80.4
Wastewater discharge quantity per 10,000 yuan of productivity	m³	70.72	64.18	60.6	55.4	49.6
Industrial wastewater treatment rate	%	79.37	84.22	89.1	90.9	90.4
Rate of treated industrial wastewater in compliance with standards	%	61.52	64.56	81	83.7	86.4
Rate of industrial solid waste used	%	59.36	63.42	72.4	77.4	74.1
Rate of industrial solid waste treated and disposed	%	59.59	63.62	72.9	78.1	74.6
Urban area under ambient noise control program	%	22.06	22.40	30.7	39.3	43.9
Items Related to Development of Urban Infrastructure						
Rate of treatment of urban garbage	%	7.944	52.71	42.1	42.2	43.7
Rate of household use of gas in city	%	85.11	96.98	86.6	89.8	91.7
Rate of households with centralized heating	%	27.73	30.95	32.2	34	34.4
Rate of green space coverage in existing urban area	%	28.14	30.4	30.7	31.4	32.4
Rate of municipal wastewater treatment	%	6.632	1.234	3.1	10.5	20.3

[a]Units of measure: μg/m³ is micrograms per cubic meter, μg/l is micrograms per liter, dB is decibels, and m³ is cubic meters, and m² is square meters.

Source: Urban Environmental Management Division, State Environmental Protection Administration.

Table 5-5. Integrated Urban Environmental Quality Index for Shanghai, 1988–1990

Item examined	*Unit of measure*[a]	*1988*	*1989*	*1990*
Items Related to Environmental Quality				
Total suspended particulates, annual average	μg/m^3	300	315	358
Sulfur dioxide, annual average	μg/m^3	110	99	95
Drinking water source compliance rate	%	85.7	85	78.6
Average ambient surface water chemical oxygen demand level	μg/l	687	717	761
Average ambient noise level in the urban area	dB	62.6	63.4	65.9
Average noise level at major roads in the urban area	dB	75.8	75.1	75.9
Items Related to Pollution Control				
Urban area enforcing smoke and dust control zone program	%	>95	91.9	98.3
Household use of briquettes	%	67.4	50.5	79.1
Compliance of industrial air emissions	%		55.9	55
Compliance of automobile air emissions	%		37.7	95
Wastewater discharge quantity per 10,000 yuan of productivity	m^3	153	129.5	115.3
Industrial wastewater treatment rate	%	27	53.5	59.4
Rate of treated industrial wastewater in compliance with standards	%	76	80.2	78.8
Rate of industrial solid waste used	%	81.4	85.3	91.3
Rate of industrial solid waste treated and disposed of	%	16	13.5	18.7
Items Related to Development of Urban Infrastructure				
Rate of household use of gas in the city	%	54.5	56.6	60.5
Rate of households with centralized heating	%			
Rate of municipal wastewater treatment	%	10.6	23.4	22.6
Rate of urban garbage centrally managed	%	98	100	100
Average green space per person	m^2	0.95	0.95	1.02

[a]Units of measure: μg/m^3 is micrograms per cubic meter, μg/l is micrograms per liter, dB is decibels, m^3 is cubic meters, and m^2 is square meters.

Source: Urban Environmental Management Division, State Environmental Protection Administration.

Table 5-6. Integrated Urban Environmental Quality Index for Shanghai, 1991–1995

Item examined	*Unit of measure*[a]	*1991*	*1992*	*1993*	*1994*	*1995*
Items Related to Environmental Quality						
Total suspended particulates, annual average	μg/m³	324	333	292	810	264
Sulfur dioxide, annual average	μg/m³	106	93	89	73	53
Drinking water source compliance rate	%	85.71	85.71	92.86	90.84	90.46
Ambient surface water chemical oxygen demand level	μg/l	557.3	629.3	592	463	458
Average ambient noise level in the urban area	dB	60.4	61.5	61.1	60.3	60.1
Average noise level at major roads in the urban area	dB	74.6	75.6	72.6	71.8	71.7
Items Related to Pollution Control						
Urban area enforcing smoke and dust control zone program	%	98.27	98.27	98.27	100	100
Household use of briquettes	%	76.85	70.04	80.04	90.13	91.48
Compliance of industrial air emissions	%	83.39	82.32	96.46	90.05	91.67
Compliance of automobile air emissions	%	70.96	65.37	80.02	79.99	80.12
Wastewater discharge quantity per 10,000 yuan of productivity	m³	90.13	84.51	66.81	56.91	46.37
Industrial wastewater treatment rate	%	76.37	83.33	79.8	82.79	81.02
Rate of treated industrial waste-water in compliance with standards	%	85.52	86.59	79.54	92.8	93.67
Rate of industrial solid waste used	%	81.33	79.2	85.1	86.23	89.85
Rate of industrial solid waste treated and disposed of	%	96.09	87.7	98.53	97.72	98.21
Urban area under ambient noise control program	%	12.27	24.08	30.81	29.11	42.15
Items Related to Development of Urban Infrastructure						
Rate of treatment of urban garbage	%	38.92	43.28	71.58	71.55	90.33
Rate of household use of gas in city	%	64.24	70.94	78.85	84.55	86
Rate of households with centralized heating	%					
Rate of green space coverage in existing urban area	%	13.96	14.47	15.11	15.71	16
Rate of municipal wastewater treatment	%	24.56	17.95	24.27	33.23	40.5

[a]Units of measure: μg/m³ is micrograms per cubic meter, μg/l is micrograms per liter, dB is decibels, and m³ is cubic meters.

Source: Urban Environmental Management Division, State Environmental Protection Administration.

treatment of wastewater has increased from about 10% to 40% in Shanghai and from 7% to 20% in Beijing. The same can be said for household use of gas for cooking. This has increased from about 55% to 86% in Shanghai, while it has stayed near 90% in Beijing. Fourth, both cities have made substantial progress in several of the pollution control indicators. For example, the treatment of industrial wastewater has risen from 27% to 81% in Shanghai and from 30% to 90% in Beijing. The percentage of treated industrial wastewater meeting discharge standards has increased from 76% to 93% in Shanghai and from 78% to 86% in Beijing.

Why, however, has ambient environmental quality showed so little improvement in these cities, particularly Beijing, despite what appears to be significant progress in expanding urban environmental infrastructure and in controlling pollution? Senior officials in both EPBs had a simple answer to this question. They said improvements in urban environmental infrastructure and in pollution control have been overwhelmed by increases in the scale of economic activity. When asked about this phenomenon, EPB officials in Changzhou, Nanjing, and Tianjin concurred with their colleagues in Beijing and Shanghai.

The phenomenon of scale effects overwhelming improvements in urban environmental infrastructure and in pollution control appears to be particularly severe in Beijing. EPB officials there said that in the past decade the city's population has increased by more than 1 million, its economic output has more than doubled, and its annual coal consumption has increased by 6 million tons. On top of this, the number of vehicles on the city's roads has increased from about 300,000 a decade ago to more than 1.3 million today. Beijing is feeling the effects of scale in other ways. Not only is it making less progress on ambient environmental quality than cities such as Shanghai, but its UEQES score and rank also have been falling. A number of years ago, it was in second place; now it is in tenth place.

Despite this decline in score and rank, officials of the Beijing EPB are optimistic about the future because citizens are demanding a cleaner environment.[49] They also are optimistic because, before UEQES, they had great difficulty affecting either environmental management (pollution control) or the level of development of urban environmental infrastructure. UEQES has enabled them to do both. For example, before UEQES they could not get the UCB to reduce the dust levels associated with urban construction. After UEQES was established and implemented in Beijing, however, the UCB developed a new standard for setting up and operating construction sites, which has significantly reduced the associated dust. Finally, they are optimistic because the mayor of Beijing is now more actively involved in environmental management. Once he discovered that Beijing was losing points in the UEQES index and dropping in rank, he agreed to implement tougher emissions standards for cars and to convert city-owned buses to use natural gas as fuel. He also began to support expansion of the subway and bus systems and the building of a light rail system.

Conclusions

There is little doubt that China's major cities remain among the most polluted in the world (World Bank 1997b, 6–13). Air pollution in most major cities is two to five times the standards set by the World Health Organization (as measured by TSP and SO_2 levels). Pollution of the country's major rivers is equally bad. About 40% of the monitored sections of rivers in the country do not meet even minimum water quality standards. Nearly half of monitored sections of rivers that run through urban areas of China are not suitable for irrigation, and only about 8% of monitored sections in urban areas meet standards for direct human contact. Some rivers, such as the Xiaoqing River in Shandong Province, are nearly dead, are devoid of fish and plant life, and barely contain any dissolved oxygen.

Critics of China's poor environmental record attribute this lackluster performance to a bargaining model of policy implementation, particularly within cities, that places too much emphasis on growth and development and too little emphasis on the environment (Lieberthal n. d.; Smil 1993; Vermeer 1998; Jahiel 1998). Critics also contend that real environmental improvement depends on more actively involving the public, on getting mayors and others in vaunted local economic bureaucracies to take environmental concerns more seriously, and on developing a legal framework that supports monitoring and enforcing the country's stringent environmental standards.[50] Although there is some truth to these allegations and proposed remedies, they are half-truths at best.

For one, there is a growing body of evidence that the country and several of China's major cities have, in important ways, delinked growth from environmental degradation. As was mentioned above, at the national level, the TSP intensity of GDP fell from 34.3 tons of TSP per million yuan in 1981 to 9.8 tons of TSP per million yuan in 1993. Overall, the TSP pollution load fell slightly, from 15.2 million tons in 1981 to 14.1 million tons in 1993; real GDP increased from 442 billion yuan in 1981 to 1,445 billion yuan in 1993. This is a remarkable accomplishment by any standard.[51] There has been a similar, if less spectacular, improvement in the SO_2 intensity of GDP.[52] Ambient TSP levels in large cities in north China fell from roughly 800 $\mu g/m^3$ in 1986 to 500 $\mu g/m^3$ in 1993. Large cities in south China fell from 600 $\mu g/m^3$ in 1986 to 400 $\mu g/m^3$ in 1993 (Wang and Liu 1999, 381). The data suggest that some provinces are experiencing similar declines in the water pollution intensity of their output, as measured by the kilograms of COD per 10,000 yuan of gross provincial product (World Bank 1997b,

60). The problem, of course, is that declines in pollution intensity have not been large enough to achieve ambient environmental quality anywhere near that currently experienced by the countries that belong to the Organisation for Economic Co-operation and Development.

The evidence on declining pollution intensities, nationally and within cities, and on moderately improving ambient environmental quality, at least in some cities, requires some explanation. Because studies of China's environmental policy instruments have found them to be largely ineffective, this leaves open the possibility that UEQES and the target responsibility system to which it is tied may explain the limited environmental successes achieved to date. Officials of the five EPBs interviewed are convinced that this is so. They cite several reasons why UEQES and the target responsibility system are making an environmental difference. First, both of these programs have captured and held the interest of mayors. As was stated above, mayors want to know why their cities rank lower than other cities in the UEQES index. They want to know how UEQES works. They want to know what they can do at what cost to improve their city's rank and score on the UEQES index. Their desires for answers to these questions are undoubtedly heightened by the fact that they now sign annual and five-year environmental responsibility contracts with provincial governors that are based on individual UEQES indicators.

Because of their keen interest in UEQES and the target responsibility system, mayors have taken the lead in getting all sections of urban government, including those in vaunted economic organizations, to focus on improving their city's score and rank on the UEQES index. This is the second reason cited by officials in EPBs for why the UEQES process works. As they stated, before UEQES, the EPB was the only organization in a city responsible for the environment. Now every organization is involved. Because virtually every sector in city government signs a contract with the mayor for the environmental target responsibility system, each organization has to devise a plan for how they will contribute to a better score on the UEQES index. These contracts serve as the basis for the mayor's target responsibility contract with the provincial governor.

The performance of a sector relative to its target is evaluated two times a year. When performance falls short of targets, sectors must explain why. The USB, EPB, utility company, and heat supply company may all have to explain to the mayor why the percentage of municipal wastewater treated is not rising or is not rising fast enough to improve the city's score on the wastewater treatment indicators in UEQES. The UCB may have to explain to the mayor why the dust levels associated with urban construction are not falling or are not falling fast enough to improve the city's score on the TSP indicator in UEQES. The ECs and IBs may have to explain why industrial wastewater treatment rates are not rising or not rising fast enough and why the percentage of industrial wastewater discharge meeting emissions standards is not rising or not rising fast enough. The PC may have to explain to the mayor why urban environmental infrastructure investment is not rising or is not rising fast enough and why pollution control investments are not rising or are not rising fast enough. And the EPBs may have to explain to the mayor and to those who have made investments in urban environmental infrastructure or in pollution control why ambient environmental indicators are not improving or are not improving fast enough.[53] One other piece of evidence suggests that mayors and those in sectoral organizations take this whole process seriously. At least part of mayors' annual personnel evaluations of sector heads is based on the environmental performance of sectors relative to targets on particular UEQES indicators.[54]

The third reason EPB officials cite for why they believe UEQES works is that it fosters an integrated approach to urban environmental management. Before UEQES, there was little, if any, coordinated discussion of environmental issues within cities. The UEQES process and the bottom-up process used by cities to set targets on each UEQES indicator have forced coordination. Much of this takes place within the EPCs that have been charged with implementing UEQES. Because membership on the EPC includes senior representatives from virtually every sector in a city, each unit of government, including the EPB, now has a better understanding of what it takes to improve the score on any environmental indicator and on the UEQES index.

Despite what we contend are the major accomplishments of UEQES and the target responsibility system, we would be remiss if we did not raise four important concerns about them. First, many in China are concerned that the UEQES index treats cities unfairly. As was mentioned above, old, poor inland cities tend to score and rank lower than young, rich coastal cities. Some of this difference appears to be due to the fact that rich cities have invested more heavily in building the capacity of their EPBs. Some of it may be due to the fact that citizens, mayors, and other government officials in rich cities are more concerned about the environment than their counterparts in poor cities.[55] And some of it may be due to the fact that some of the dirtiest industrial installations, such as electroplating and textile-dyeing factories, have been forcibly moved out of large, rich coastal cities to small, poor inland cities.[56] But there are no systematic studies of these alleged differences. Moreover, it not yet clear what, if anything, should be done about these differences.

Second, there is some concern that too much effort within cities and EPBs is directed at increasing a city's score on particular indicators and on the UEQES index, and that too little is directed at identifying key environmental risks and cost-effective ways to reduce them. As far as we know, there has been virtually no research on this issue. But if it turns out that UEQES and the target responsibility system do in fact divert attention away from serious environmental health risks and ways to reduce them, neither may be a particularly effective long-run tool of urban environmental management.

Third, all too little is known about whether UEQES and the bargaining model of environmental policy implementation on which it depends can ultimately achieve the kinds of improvements in ambient environmental quality that China needs to make. The evidence on other areas of policy implementation in China suggests that bargaining approaches contribute to various pathologies, such as failed implementation, stalemate, and minority veto (Lampton 1992, 37). If these problems seriously afflict implementation of UEQES, it may not be a particularly effective long-run tool of environmental management.[57] Said another way, it is not yet clear whether UEQES is a viable long-run model for the integrated environmental management of cities or simply an effective stopgap measure that will give way once China decides to implement a more traditional, stringent monitoring and enforcement program.

Fourth, all too little is known about how China's implementation of an integrated approach to urban environmental management with public disclosure, as is manifested by UEQES, compares with what cities in other parts of the world are doing. Both Indonesia and the Philippines have sustainable city programs. In addition, as was described in Chapter 4, Indonesia has a public disclosure program. It could be quite interesting to initiate exchanges among officials of these cities as well as to compare their performance. Much might be learned from this that would be useful for all.

Notes

[1]At the United Nations Conference on the Human Environment in Stockholm in 1972, the Chinese insisted that the primary responsibility for pollution rested with the rich capitalist economies, and they defended the right of developing countries to exploit their own resources without interference from outside parties (Ross 1988, 137).

[2]Even though China has been participating in the Intergovernmental Panel on Climate Change and is a signatory to the Montreal Protocol, its action in each area is dominated by domestic considerations. For example, it projects chlorofluorocarbon production to increase from 60,000 tons in 1996 to 160,000 tons in 2010, and it has not assigned an official organization to coordinate ozone protection issues (World Bank 1992a, vol. II, 94–96).

[3]For a discussion of the implications of decentralization and liberalization, see Breslin (1996), Davis et al. (1995), and McElroy et al. (1998).

[4]As early as 1979, popular protests over industrial pollution in Shanghai led the city government to partially or completely close down 49 factories (Ross 1988, 146).

[5]For example, after a local shipyard in Wuhan failed again and again to reduce coal dust, while coal was being unloaded one day, an angry mob broke into the shipyard and smashed machinery. Subsequently, the party criticized the leaders of the Ship

Administrative Bureau for neglecting the people's interest, and the vice premier of the country forced an investigation of the incident (Ross 1988, 147).

[6]A 1984 study indicated that pollution cost China 9% of the gross value of agricultural and industrial output in 1982 (Ross 1988, 135–136). The human health costs of pollution have also been high. The incidence of lung cancer in Shanghai rose from 5.25 cases per 100,000 inhabitants in 1960 to 35 cases per 100,000 inhabitants in 1976, and there are numerous instances of mercury poisoning (Ross 1988, 136).

[7]Bureaucratic politics dominates domestic politics because of the totalitarian nature of the Chinese state, which has limited the emergence of civil society and the role of independent groups in civil society in politics and policymaking (Okabe 1998; McCormick et al. 1992).

[8]The government' s most successful pollution reduction programs are a pollution levy for wastewater emissions (Wang and Wheeler 1999) and a unique nationwide rating, ranking, and public disclosure program that annually assesses the environmental performance of the country's major cities. For reasons that will become obvious, the latter is the focus of this chapter.

[9]A water pollution levy was first introduced in Suzhu in 1979 and subsequently extended to all of China starting in 1981; and a water discharge permit system was first tried in 17 cities in 1987, expanded to 107 cities in 1990 and to 260 cities in 1992, and applied nationwide in 1995 (Panayotou 1998, 434).

[10]As Jahiel states (1998, 759), there is great variability in the funding, staffing, and work methods of local environmental agencies. Those in rich coastal provinces and big cities are better funded, have more trained technical staff, and have better enforcement practices than their counterparts in poor inland areas.

[11]The environmental responsibility system made local political officials responsible for environmental outcomes within their jurisdiction to officials one level higher up in China's bureaucratic hierarchy. Thus mayors became responsible for the environmental performance of cities to provincial governors; and governors, in turn, became responsible for the environmental performance of provinces. Both were held accountable for environmental performance through the signing of environmental performance contracts with their immediate superiors (Jahiel 1998, 777).

[12]Limited time treatment was adopted to signal to factories that environmental authorities were responsible for ensuring that wastes were controlled within a limited time frame. If deadlines for treatment were not met, environmental authorities could impose large penalties. Centralized pollution control was a response to the particular needs of small and medium-sized enterprises (Jahiel 1997, 83).

[13]China has a unique, multilayered bureaucratic structure with territorial divisions that stretch from the national government in Beijing to provinces, cities, counties, townships, and villages. Each national agency (ministry) has its counterpart at each level of government. In addition, each local level of government (province, city, county, and township) has its own bureaucratic structure. As a result, local representatives of national agencies (ministries) serve two masters—both their national parent and local government (provincial governor for provinces or mayor for cities).

[14]For a discussion of China' s bureaucratic ranking system and its implications for environmental policy, see Lieberthal (n.d.) and Jahiel (1998).

[15]Vermeer (1998, 955) has argued that low salaries and salary erosion have made it difficult for EPBs and EPOs to attract and keep high-quality staff. Panayotou (1998, 461) has argued that the dependence of EPBs and EPOs on revenues from the pollution levy discourages them from pushing firms to comply with the regulations, for fear that this would undermine an important source of financial support.

[16]Depending on the bureaucratic ranking of SEPA or its predecessor, NEPA, the bargaining relationship between EPBs and UCBs can be either vertical or horizontal.

[17]Even though China's hierarchical and functional bureaucratic structures reach down to the lowest level of government, local representatives of these central functional bureaucracies depend on local governments for funding, salaries, and many perquisites. Because of this, they serve two masters—the local authorities where they work and live, and the central authorities in the functional bureaucracy in which they work. But because local representatives of central functional bureaucracies are financially beholden to local government, this means that central authorities in functional bureaucracies cannot normally order their local counterparts to implement new policies. They can, at best, negotiate with them over implementation.

[18]City-level finance and tax bureaus "whip the fast oxen" by heavily taxing profitable enterprises while subsidizing those facing poor objective conditions (Walder 1992, 319).

[19]Bargaining is also more likely to occur when central government leaders disagree over a new policy, when they fail to give a policy issue a high priority, and when it is difficult to measure compliance with a new policy initiative (Lampton 1992, 34–35).

[20]In 1989, provinces made their own assessments of 175 cities, and by 1990 this had increased to 242 cities (Sinkule and Ortolano 1995, 36). By 1997, 570 cities were being evaluated (communication with SEPA, January 20, 1999).

[21]Both Ren Min Ri Bao and China Central Television publicize the results of UEQES (comments from interviews with officials of the Beijing EPB).

[22]For example, see Panayotou (1998), Spofford et al. (1996), Sinkule and Ortolano (1995), and Ross (1988). For an exception that focuses on wastewater emissions by factories, see Wang and Wheeler (1996, 1999).

[23]For discussions of the bargaining model of policy implementation, including environmental policy implementation, see Lieberthal (n.d., 1992), Lampton (1992), and Walder (1992).

[24]EPB officials in Tianjin noted that—despite SEPA' s tough energy efficiency standards in its sustainable cities index—the city was committed to becoming sustainable.

[25]For example, the index in Changzhou includes parameters for floating debris in the Grand Canal and for the number of garbage dumps near the canal. In addition, some northern cities have added the centralized heating rate as a parameter; and others, such as Zhejiang in the south, which has a large number of water courses, include the treatment rate of water courses as a parameter.

[26]Because weights add to 100, the TSP indicator accounts for 4% of the total weight in the current UEQES index.

[27]$TSP = 4^{*}(0.6 - 0.6)/0.42 = 0$.

[28]$TSP = 4^{*}(0.6 - 0.18)/0.42 = 4$.

[29]The data on ambient air quality are from Wang and Lui (1999, 381).

[30]Thus, as a city takes steps to improve ambient TSP concentrations, SEPA can ratchet up performance requirements by, for example, reducing the maximum TSP for northern cities to 500 μg/m^3 and lowering the minimum to 100 μg/m^3. Assuming no change in the weight for the TSP indicator, the new TSP formula becomes TSP = $4*(0.5 - X)/0.4$, where if X = 100 μg/m^3 (0.1), the score on this indicator will be TSP = $4*(0.5 - 0.1)/0.4 = 4$.

[31]This comment is based on discussions with senior EPB officials in Beijing.

[32]The current minimum value for percentage of sewage treated is zero.

[33]If X is 30% or less, a city's score on this indicator is 0.

[34]This characterization is far from perfect. Note that Chengdu, a poor inland city, scores higher and ranks higher than Ningbo, a rich coastal city.

[35]The next section of this chapter addresses an even more important question: Is there any evidence to suggest that city governments use the UEQES examination process to improve urban environmental outcomes?

[36]This discussion is based on interviews with officials in one city-level EPB.

[37]This appears to be similar to the effect of public disclosure on the pollution behavior of the chemical industry in the United States and of heavy water-polluting industries in Indonesia. In both instances, public disclosure of toxic emissions (in the United States) and of loads of biological oxygen demand and chemical oxygen demand (in Indonesia) galvanized senior corporate managers to take action to significantly reduce pollution (Arora and Cason 1995; Afsah and Vincent 2000).

[38]Based on comments made by EPB officials in Changzhou.

[39]Based on comments made by EPB officials in Tianjin.

[40]The UEQES examination process and the target responsibility system are two of the nine major environmental policy instruments available to SEPA and local EPBs. The others are EIAs, the "three simultaneous" program, the pollution levy, the discharge permit system, limited-time treatment, centralized pollution control, and total load control (Sinkule and Ortolano 1995, 25; discussions with SEPA officials, winter 1998).

[41]As stated by Ross (1988, 37), reward and punishment typically take the form of positive or negative publicity, particularly when the results are announced in the newspaper and on television.

[42]Mayors also sign annual environmental performance contracts with provincial governors. The description that follows is based on discussions with EPB officials in Nanjing (interviews, winter 1998).

[43]The EC will also sign an annual performance contract with the mayor.

[44]This comment was made by EPB officials in Changzhou, Nanjing, and Shanghai (interviews, winter 1998).

[45]What follows is based on discussions with EPB officials in Tianjin (interviews, winter 1998).

[46]Based on a discussion with EPB officials in Nanjing (interviews, winter 1998) .

[47]What follows is based on discussions with EPB officials in Changzhou (interviews, winter 1998).

[48]This also required the installation of pollution control equipment by a chemical factory near the new intake. The equipment cost more than 900,000 yuan. The local EPB used its pollution levy fund to cover 20% of the cost of the equipment.

[49]Two manifestations of their demands: Citizen complaints to the Beijing EPB hotline have been increasing. And the Beijing EPB now reports ambient air quality weekly, and citizens watch this index closely.

[50]For example, the World Bank (1997b, 59) contends that the poor environmental record is a consequence of insufficient public involvement in environmental issues, excessive local government interference to protect enterprises, weak environmental monitoring, and lenient enforcement.

[51]If TSP emissions per unit of GDP had not changed, total emissions of TSP in 1993 would have been more than 45 million tons. Actual emissions in 1993 were only 14.16 million tons (Wang and Lui 1999, 385).

[52]The sulfur dioxide intensity of GDP declined from 19.9 tons per million yuan in 1981 to 12.4 tons per million yuan in 1993 (Wang and Liu 1999, 385).

[53]Officials of one EPB said that before UEQES, virtually no one in the city paid attention to the EPB's monitoring data. Some in the West went so far as to suggest that these monitoring data had been faked. Now everyone pays attention to monitoring data. If the Urban Services Bureau increases the percentage of municipal wastewater treated, they want to see improvements in the ambient environmental quality of rivers. If this does not happen, they ask the EPB why. If plants in an industry attached to an industrial bureau install air pollution equipment, they want to see improvements in ambient air quality in their neighborhoods. If this does not happen, they also turn to the EPB for an explanation of why (interviews with officials of the Beijing EPB, winter 1998).

[54]Senior officials of several EPBs stated that this was now standard practice in their city (interviews, winter 1998).

[55]Spofford et al. (1996, 3–15) argue that governments of poor cities such as Chongqing are more likely than their counterparts in rich cities to short-circuit environmental review of new investments. This happened recently in Chongqing when the government eliminated the third and final review of new investment by the local EPB in the "three simultaneous" system.

[56]EPB officials in Chongqing have stated that they cannot keep up with rising pollution because heavily polluting factories that were forced to move from large, rich coastal cities have been relocating in Chongqing (Spofford et al. 1996, 3–19).

[57]There is some evidence that this may be happening. As representatives in one EPB stated, everything depends on the city's financing actions to improve scores on individual indicators in UEQES. For example, if the city does not include plant-level investment in pollution control in the city's investment budget, the plant will not make the investment (interviews with officials of the Tianjin EPB, winter 1998). Sometimes, EPBs find this frustrating.

6

Pollution Management in Malaysia and Thailand

Malaysia and Thailand are resource-rich, export-oriented economies. Forty years ago, both economies produced and exported primary commodities. Malaysia specialized in rubber and tin; Thailand, in rice, tin, teak, and rubber (Jomo K.S. 1986; Ingram 1971). Following successful, if short-lived (particularly in Malaysia), import-substitution industrialization programs that helped diversify their economies, both turned to the promotion of foreign investment and the exporting of manufactures (Lee 1986; Rock 2000d). The combination of successful import-substitution and export-oriented industrialization programs facilitated rapid expansion in manufactures, manufactured exports, and industrial pollution loads.[1]

The governments of both countries responded early to the rising industrial pollution loads and deteriorating ambient environmental quality attending high-speed industrial growth. Thailand's Factory Act of 1969 made the Ministry of Industry (MOI) responsible for setting and enforcing emissions standards (O'Connor 1994, 69, 100). In 1975, the Thai parliament passed the Improvement and Conservation of the National Environmental Quality Act (NEQA; hereafter, later versions of this law are identified with the year given after "NEQA"). NEQA created an environmental oversight agency—the Office of the National Environmental Board (ONEB)—and initiated environmental impact assessments (EIAs). Landmark environmental legislation in Malaysia was promulgated in 1974, when the government enacted the Environmental Quality Act of 1974 (EQA 1974) (Sani 1993, 71–74). This act complemented several sector-specific acts affecting the environment and led to the establishment of the Department of the Environment (DOE), which is responsible for monitoring and enforcement.[2] By 1979, the Malaysian government had promulgated nine different environmental regulations covering air and water emissions.

Despite the broad similarities in political economies and in quick, early legislative action to protect the environment during rapid industrial growth, the evolution of industrial pollution management in Malaysia and Thailand took decidedly different paths. Whereas Thailand followed NEQA with several other pieces of important environmental legislation, including a new environmental framework law (NEQA 1992), it has not created the legislative or institutional framework to successfully monitor and enforce emissions regulations on point sources of pollution (World Bank 1994b, 12). Because of this, ambient environmental quality in Thailand is poor, and Thailand has the dubious distinction of being the only East Asian newly industrializing economy (NIE) without at least one important industrial pollution management success (Kanittha 1996).[3] Malaysia, conversely (much like Singapore, its richer neighbor to the south), has successfully demonstrated that it is possible to clean up the environment while experiencing rapid industrial development. This has been particularly true of the government's highly successful efforts to delink water pollution from crude palm oil (CPO) production and palm oil exports (Vincent 1993). But it is also at least partially true for air pollution from point sources. Because of these successes, ambient air and water quality in Malaysia is improving with time.[4]

The Evolution of Pollution Management

Before turning to an explanation of the differences in the evolution of industrial pollution management in Malaysia and Thailand, it is important to describe and assess both governments' responses to rising industrial pollution loads in more detail. We first turn our attention to Thailand, and then to Malaysia.

Thailand

Thailand has an admirable environmental legislative record (Limanon 1999, 20, 29, 31–32). The Factory Act of 1969 (as amended up through 1992) vests authority and responsibility for setting emissions standards and enforcing them with the MOI. The Industrial Estate Authority of Thailand (IEAT) Act of 1979 vests the IEAT with the authority to monitor and enforce emissions standards within industrial estates and to build and operate common wastewater treatment facilities within estates. The Electricity Generating Authority of Thailand (EGAT) has the responsibility to develop environmental policies covering emissions from its generating plants.[5] NEQA 1975 complemented these activities by creating a high-level environmental oversight agency, the ONEB. Following this, the National Economic and Social Development Board's Seventh Plan (1992–1996) established specific environmental improvement targets for the first time. To meet those targets, the government integrated environmental criteria into Board of Investment promotional privileges and revised the NEQA 1975 in 1992.

NEQA 1992 is Thailand's first comprehensive environmental framework law. (This and the next paragraph, unless noted, are taken from Limanon 1999, 22–27.) Under NEQA 1992, the National Environmental Board (NEB) was elevated to subcabinet status and given the responsibility to oversee the implementation of the law. In addition, NEQA 1992 restructured the Ministry of Science, Technology, and the Environment (MOSTE) by establishing three new offices. A Pollution Control Department (PCD) was established with the responsibility to set emissions standards and coordinate the monitoring and enforcing of those standards with responsible agencies. A separate Office of Environmental Policy and Planning was given the responsibility to develop a national environmental plan and to administer EIAs. A Department of Environmental Quality Promotion was created to enhance public education on environmental protection, particularly with the media, as well as to provide specialized environmental knowledge to other government agencies.

NEQA 1992 appears to provide for a strong command-and-control regulatory framework by granting the PCD the authority to set standards and to monitor and enforce them. Although initial authority over point sources of pollution (factories) still lies with the Department of Industrial Works (DIW) in the MOI, the PCD can impose tougher standards on polluters. The PCD also has the authority to require plants to install pollution control equipment, and it can fine them up to four times the daily cost of operating such equipment (Sunee and Canino 1998).

In addition, NEQA 1992 holds polluters to strict civil liabilities and criminal penalties, even if owners of polluting facilities did not willfully pollute. NEQA 1992 enhances nongovernmental organization (NGO) participation in environmental matters by granting them legal standing in the courts, and it legally institutionalizes a citizen's right to know. NEQA 1992 also mandates EIAs for certain new economic activities. Under the legislation, a committee of technical experts reviews private-sector projects subject to EIAs, whereas the NEB reviews government projects subject to EIAs. In 1997, the government went so far as to amend the constitution so that citizen participation was required for all EIAs. Taken together, the creation of a command-and-control, standard-setting, monitoring-and-enforcement office within

MOSTE, alongside clear and rigorous requirements for EIAs and of a citizen's right to know, suggest that Thailand has gone a long way toward more effectively regulating industrial pollution.

All of this has not worked very well in practice, however. Overlapping responsibilities for monitoring and enforcement, alongside limited resources to carry out these tasks, has meant that both have been weak. Despite the creation of a central PCD in MOSTE, monitoring and enforcement responsibilities remain divided among numerous agencies. EGAT enforces emissions standards on its generating plants. IEAT monitors and enforces standards within industrial estates and builds and operates centralized wastewater treatment facilities within estates. Although this kind of delegation of environmental authority to economic actors conceivably could work, in practice EGAT has dealt with excessive sulfur dioxide emissions by shedding loads rather than by installing pollution control equipment (World Bank 2000a, 5). For its part, IEAT has delegated authority to individual estate managers. Because these managers are also charged with attracting and keeping clients, delegating authority to them appears to involve real conflicts of interest (Limanon 1999, 51).

It is not surprising that most industrial estates have poor environmental records. A 1994 study of the environmental practices of 49 estates revealed that most were not complying with EIA requirements (MOSTE 1997). Sixty percent of the IEAT-managed estates were out of compliance with emission standards, most did not adequately monitor wastewater effluent, and most used unreliable contractors to handle hazardous waste (Limanon 1999, 55). A followup study by the PCD in 1998 that looked at 11 IEAT-managed estates and 13 private estates suggests that there was little improvement (MOSTE 1998; Limanon 1999, 56–57). This study found that 65% of estates had significant odor and dust fallout problems, most still did not adequately treat or dispose of hazardous waste, and 43% were out of compliance with wastewater effluent regulations. In one large "showcase" estate, Map Ta Phut, on the Eastern Seaboard, air emissions more than once caused students in a nearby school to choke, vomit, and faint. The government finally addressed this problem not by requiring factories to reduce their pollution but rather by relocating the school 5 kilometers from the offending estate (Kanittha 1997, 2).

Monitoring and enforcing emissions regulations by the DIW is little better than that by IEAT or EGAT (O'Connor 1994, 100, 101, 107). The DIW is responsible for ensuring that emissions standards are incorporated into applications for factory licenses; in fact, licenses are often granted before plant construction begins. The DIW also has authority to inspect treatment facilities before initial operations begin and before licenses are renewed. But given the DIW's small staff and limited authority, there is not much evidence that it has been able to carry out these tasks.

Unlike pollution control agencies in Malaysia, Singapore, South Korea, and Taiwan, the PCD until recently did not collect and keep pollution-related data on inspections, compliance, and penalties. What little data it did collect suggest that the DIW's monitoring, inspection, and enforcement programs are not effective. For example, in 1995, the DIW inspected 5,213 smokestacks and 22,815 wastewater effluent streams, and issued 1,488 compliance orders and 190 factory-closing orders (Limanon 1999, 36). Because this represents a very small percentage of Thailand's industrial point-source polluters, the DIW is forced to rely on the quarterly emissions reports that polluters submit to it.[6] Plant reporting is sporadic at best. And of those reports submitted, almost all show that plants are in full compliance. Given poor ambient air and water quality, these reports can hardly be accurate (World Bank 1994b, 146).

To make matters worse, Thailand lacks an emissions permit system, and the PCD continues to lack full legal authority to inspect and enforce its emissions standards (Limanon 1999, 37). If the PCD suspects factories of violating standards, it must ask the DIW to act. If the DIW fails to respond, the PCD has the authority to take action. In practice, this rarely happens. As a recent World Bank report on industrial pollution management in Thailand states:

> Public institutions responsible for environmental ... management are highly segmented with little coordination among them. Compliance and enforcement functions are split among the Departments of Industrial Works, Pollution Control, and Land Transport. This has led to poor implementation of the National Environmental Quality Act.... [Moreover] current enforcement efforts suffer from inadequate procedures, institutional overlap, insufficient staff capacity, a lack of incentives, and weak monitoring and reporting capability. (World Bank 2000b, 16)

It is not surprising that the report concluded that noncompliance is widespread. As a result, ambient environmental quality is poor (Kanittha 1996, 1–3).

Because Thailand's command-and-control regulatory institutions are so weak and ineffectual, some in government have tried to leapfrog to new policy initiatives that build on internal and external pressures to get industrial facilities to clean up their environment (interviews in Thailand with a PCD official, 1996). With respect to the former, the government has enshrined a community's right to know in its most recent constitution, and it is working hard to build the capabilities of indigenous environmental NGOs. But as will be demonstrated in this chapter, this effort has not been very successful. The government is also relying heavily on the most recent standards from the International Standards Organization (ISO; the standards are known as "ISO 14000") to provide a stimulus to environmental cleanup.

The Thailand Industrial Standards Institute (TISI) within the MOI has been very active on the ISO's Technical Committee 207 (TC 207), which developed the criteria for ISO 14000 certification. TISI has representatives on each TC 207 subcommittee, actively trains environmental auditors, and has a pilot project to prepare 10 large companies for ISO 14000 certification (interviews in Thailand with a TISI official, 1996).[7] But so far, this effort does not seem to be having much effect. As of December 2000, 302 companies in Thailand were ISO 14000 certified; most are multinationals, and all but 24 were certified by others rather than by TISI (TISI 2000). The limited success of the government's efforts to employ internal and external pressure on polluters suggests that this tactic may not work in the absence of a more effective command-and-control regulatory agency. This is consistent with the experiences of the countries that belong to the Organisation for Economic Co-operation and Development, where these new initiatives depend on information supplied by traditional command-and-control environmental agencies.

Malaysia

As was the case in Thailand, the government of Malaysia began addressing pollution in the early stages of its industrial revolution (Sani 1993, 71–73). In 1974, the government passed comprehensive environmental legislation, the EQA. The act created an Environmental Quality Council to advise the minister in charge of the environment. The act contains provisions governing, among other things, air (section 22), noise (23), land (24), and water pollution (27). Unlike Thailand's NEQA 1992, Malaysia's EQA 1974 grants the director-general of the environment the right to limit discharges from polluting sources and the right to require polluters to install, operate, repair, or replace pollution control equipment. In 1975, DOE was created to assist the director-general in implementing regulations.

Ultimately, DOE was made responsible for monitoring environmental quality, estimating the environmental effects of new projects, and designing and enforcing environmental regulations (Vincent et al. 1997, 26). DOE has also been authorized to prescribe particular classes of industrial premises and to require them to obtain licenses, with explicit environmental conditions, to operate. These environmental conditions are to take account of whether it is practical to adapt existing equipment for pollution control. In addition, they are to be imposed in light of the economic life of existing equipment and the cost incurred in getting factories to comply with environmental regulations (Vincent 1993, 7).

DOE also has the authority to vary license fees according to the class of premises, location of premises, the quantity of pollutants, and the class of pollutants (Vincent 1993, 7). After the passage of EQA 1974, the central economic agencies began demonstrating that they also would take the environment more seriously. The Third Malaysia Plan (1976–1980) stated that environmental improvement would receive the full attention of government (Vincent et al. 1997, 23). The Fifth Malaysia Plan (1986–1990) spelled out the government's guiding environmental principles. They included maintaining a clean and healthy environment, balancing socioeconomic development against the need to maintain sound environmental conditions, and incorporating environmental considerations into project planning (Sani 1993, 71).

Within five years of EQA 1974, DOE had promulgated 9 new environmental regulations and orders affecting industry (Sani 1993, 74). By 1989, 15 different regulations and orders governed pollution control in Malaysia. The chief regulations and orders include (World Bank 1993, 15):

- Environmental (Prescribed Premises) Order for CPO in 1977
- Environmental (Prescribed Premises) Regulations for CPO in 1977
- Environmental Quality (Licensing) Regulations in 1977
- Environmental (Prescribed Premises) for Raw Natural Rubber Order in 1978
- Environmental (Prescribed Premises) for Raw Natural Rubber Regulations in 1978
- Environmental Quality (Clean Air) Regulations in 1978
- Environmental Quality (Compounding of Offences) Regulations in 1978
- Environmental Quality (Sewage and Industrial Effluents) Regulations in 1979

The orders and regulations for CPO and raw natural rubber address rising pollution from oil palm and rubber processing. The clean air regulations govern industrial air emissions, whereas the sewage and industrial effluents regulations govern industrial water emissions.

Unlike the PCD in Thailand, Malaysia's DOE has a substantial number of enforcement tools. It has the authority to monitor factory emissions, inspect factories, require factories to submit independently audited emissions reports, issue fines, compound fines, and direct polluters to take remedial actions (Sani 1993, 72–73, 95). It can also issue prohibition orders and prosecute those in violation of emissions standards. Both the clean air regulations and the sewage and industrial effluents regulations permit DOE to issue contravention licenses when there are no technologically and economically practical means to meet emissions standards.

A range of evidence suggests that Malaysia's DOE has been substantially more effective than Thailand's PCD in controlling industrial pollution.[8] To begin with, compliance rates of manufacturing facilities with air and water emissions standards in Malaysia appear high. The overall compliance rate with industrial water emissions standards was 83.8% in 1992 and 86% in 1998, whereas that for industrial air emissions standards was 81.6% in 1992 and 78% in 1998 (DOE 1992, 20; 1998, 1).

The data on ambient air and water quality tell a similar story. On the basis of DOE's water quality index, only 5% of the country's rivers were rated as clean in 1980, 25% were classified as worse than slightly polluted, and 30 rivers or stretches of rivers were classified as moderately or grossly polluted (Vincent et al. 1997, 17).[9] By 1989, 49 rivers were rated clean, 34 were slightly polluted, and only 3 were seriously polluted (Sani 1993, 38). There was some deterioration in the quality of water in rivers by 1998 (33 rivers were still clean, 71 were slightly polluted, and 16 were seriously polluted). The major sources of water pollution in 1998 were agricultural runoff, domestic sewage, earthworks, and land-clearing activities (DOE 1998, 1).

Between 1989 and 1994, concentrations of suspended particulates in the air varied from 90 micrograms per cubic meter ($\mu g/m^3$) of air to slightly more than 150 $\mu g/m^3$ in industrial and heavily trafficked areas in Malaysia (DOE 1994, 48–49). By 1998, average air quality throughout the country was rated good (below 90 $\mu g/m^3$ of air), except for three weeks in March, when local forest and peat fires sparked an air quality hazards warning (DOE 1998, 1). In the capital, Kuala Lumpur, ambient air quality as measured by concentrations of suspended particulates in the air improved over time. They ranged between 155 and 172 $\mu g/m^3$ of air between 1978 and 1982, and fell to 135–139 $\mu g/m^3$ between 1982 and 1986, and to 119–144 $\mu g/m^3$ between 1987 and 1990. By 1998, suspended particulates in the air averaged 85 $\mu g/m^3$ of air (World Bank 1998, 4; 2000a, 163). It is not surprising that, as the 1997 *Environmental Quality Report* stated, "Enforcement visits remained the punch-line activity of the department" (DOE 1997, 1).

In 1997, DOE conducted 5,290 water effluent enforcement visits and 7,660 air emissions enforcement visits of industrial sources. As a result, DOE's enforcement actions in that year included 275 court prosecutions, fines totaling 2.39 million ringgit, and compounds of fines totaling 2.07 million ringgit (DOE 1997, 1; 1998, 2).

But DOE's greatest success came in the almost complete delinking of water pollution from CPO production and exports (Vincent 1993, 1). The government began promoting palm oil in 1962 as an alternative to rubber (Pletcher 1991, 627). Subsequently, the Federal Land Development Authority (FELDA) began promoting palm oil by developing new lands and providing ethnic Malay settlers with infrastructure and technical assistance to grow palm oil in nuclear estates centered on CPO-processing mills. Because palm oil was more profitable than rubber, private estate production and FELDA-financed palm oil resettlement schemes expanded rapidly. Between 1960 and 1969, those schemes grew from 54.7 thousand to 177.4 thousand hectares (ha) (Pletcher 1991, 625).

Following race riots in Kuala Lumpur in 1969, the government announced a New Economic Policy designed to reduce the incidence of poverty in Malaysia from 50% to 20% and increase indigenous Malay ownership in share capital to 30% by 1990 (Crouch 1996, 25). Because private palm oil estates were largely foreign owned, the government targeted them for takeover. By the mid-1980s, the government's Permodalan Nasional Berhad (PNB, the National Equity Corporation) had successfully wrested control of all the country's major private palm oil plantations away from foreign owners without disrupting either growth in output or exports (Pletcher 1991, 630).

Because the incidence of poverty among ethnic Malay farmers was much higher than among nonethnic Malays, the government's efforts to reduce poverty focused on increasing the incomes of ethnic Malay farmers. FELDA's resettlement schemes were an important part of the government's rural antipoverty programs. Because of this, FELDA rapidly expanded palm oil resettlement schemes as they grew from 51.4 thousand ha in 1969 to 377.6 thousand ha in 1984 (Pletcher 1991, 625). Although there is some doubt about the degree to which the FELDA schemes attracted the poorest of Malaysia's rural poor, there is little doubt that, as Pletcher states:

> Taken together, the investments in resettlement of small-holders onto FELDA schemes and the buyouts of foreign owned plantation companies represent the success of policies focusing on poor and landless peasants and of policies asserting national (albeit bourgeois) control over a lucrative and growing industry. (Pletcher 1991, 631)

Unfortunately, this success came at great environmental cost. (This and the next several paragraphs draw heavily from Vincent 1993.) Palm oil processing is relatively simple, but quite polluting. Palm oil fruit is first sterilized by steam to loosen the fruit from bunches. The fruit is then stripped from fresh fruit bunches and pressed to release the oil. The oil is clarified, separated, purified, and dried. This process requires a lot of water—1 ton of water is needed to process 1 ton of fruit. It also causes a lot of pollution—2.5 tons of palm oil effluent is created for every ton of CPO, and CPO effluent is 100 times more oxygen depleting than domestic sewage. Because palm oil fruit must be processed as soon as it ripens, most CPO-processing mills are located near the source of supply and along Malaysia's rivers and streams. Because CPO production in Malaysia grew rapidly, it quickly became the major source of water pollution.

By 1975, CPO effluent was equivalent to the amount of raw sewage generated by 10 million people, roughly equal to the entire population of Malaysia in 1975. By 1977, concentrations of palm oil effluent in rivers were so high that fish could no longer live in 42 of the country's major rivers. As oxygen levels in those rivers continued to fall, aerobic decomposition was replaced by anaerobic decomposition. This released hydrogen sulfide and other substances with putrid odors. Because rivers and streams in rural Malaysia are the primary source of drinking water and of protein for many villagers, the combination of declining fish stocks, more and more polluted freshwater, and a rising stench led to a flood of criticism. Between 1974 and 1978, effluent from CPO mills was the major source of public complaints about water pollution.

After the passage of EQA 1974 and establishment of DOE in 1975, DOE created an expert committee to investigate treatment technologies and advise it on economically sound and sensible regulations for CPO mills. Ultimately, the CPO industry and DOE settled on adapting a series of treatment ponds used elsewhere to treat the organic effluent from CPO mills. As evidence accumulated that a series of treatment ponds with waste-eating bacteria was environmentally successful and economically viable, in July 1997 DOE announced emissions standards on eight pollutants, including biological oxygen demand (BOD), chemical oxygen demand, total suspended solids, ammoniacal nitrogen, organic nitrogen, and pH. These regulations required CPO mills to apply for yearly operating licenses, and DOE announced that approval of license requests depended on the effluent treatment system proposed by applicants. Licensed CPO mills were also required to file quarterly emissions reports with DOE, and DOE announced that the standards would get tougher over time. Initially, the standards were not mandatory. Instead, DOE allowed mills to pay an excess fee based on the BOD load in effluent. But DOE hoped that it had set the fee high enough to induce compliance.

By the time the regulations took effect, palm oil effluent had grown to an amount equivalent to the domestic sewage of 15 million people, and many in the industry feared that the new environmental regulations would ruin the industry. Yet during the first year of the regulations, the daily discharge of BOD from CPO mills declined by 43%, or from 220 to 125 tons. In the second year, DOE made the regulations mandatory and tightened emissions standards. After a serious, widely publicized "pollution incident" in which a CPO mill's illegal effluent holding ponds broke, flooding and destroying a village, DOE suspended the mill's operating license and began taking legal action against other mills. As the system of anaerobic treatment ponds began to work, the BOD load from CPO mills declined from the equivalent of the domestic sewage of a population of 15.9 million to that of 2.6 million, while CPO production grew 44%. Following a survey of the abatement practices of CPO mills by the Palm Oil Research Institute of Malaysia in 1980–1981, DOE tightened emissions standards to less than 100 parts per million. By 1991, 75% of the CPO mills met this standard, and CPO effluent was less than 1% of what it had been before the onset of regulation. In short, the link between CPO production, CPO exports, and CPO effluent had been all but severed.

The Domestic Politics of Pollution Management

Why has Malaysia been so much more successful at industrial pollution management than Thailand? Broad similarities in initial conditions (both are resource-rich economies) and in industrial development strategies (both relied initially on import substitution, but both subsequently shifted to export-led development) suggest that both faced similar international economic and political pressure. Although this might have pushed both to clean up the environment, there is little evidence that international pressure from foreign investors, buyers, donors, or NGOs had much effect. This is particularly true in Malaysia, where the government routinely excoriates the international community, international capital, and donors. But it is also true of Thai governments, which have been proud of the fact that Thailand was never colonized, and which have been known to go their own way against the "best" advice of donors (Rock 1994, 24). Moreover, until recently neither Malaysia nor Thailand has taken much environmental advice or borrowed from the World Bank to build their environmental agencies.[10] And both have ignored the advice of donors and NGOs about the need for more rational forest policies.

The broad similarities between the two countries in the political economy of policymaking also make it difficult to account for their differences in industrial pollution management. Until quite recently, Malaysia and Thailand have been known for their bifurcated states—states with strong macroeconomic policymaking and weak microeconomic industrial policymaking riddled with patron–client ties (Rock 1995, 747–749; Salleh and Meyanathan 1993, 2). Because industrial pollution management is quintessential microeconomic policy, one might have expected patron–client ties between government officials and big business in Malaysia and Thailand to forestall attempts to reduce industrial pollution. This might at least partially explain the lack of effective management in Thailand, but it does not explain why Malaysia has been so much more effective at managing industrial pollution.

Because both governments are democratic, one might expect that public pressure and election cycles would induce them to clean up the environment.[11] As will be argued in this chapter, this pressure does help to explain why the government of Malaysia addressed pollution from CPO production, but it does not explain why Thai governments have not cleaned up the environment. What is so surprising about the Thai case is the apparent resistance of successive democratic governments to address industrial pollution in the face of repeated, highly publicized pollution incidents.[12] As was demonstrated in Chapter 3, democratization in Taiwan, by unleashing an environmental protest movement, proved to be a powerful engine of environmental change. Something similar appears to be at work in South Korea (Eder 1996). Why, it must be asked, has this not been the case in Thailand? As will be seen, part of the answer lies in Thailand's elite-based democratic transition, and part lies in the structure of its institutions of democratic governance. This line of reasoning suggests that the domestic politics of industrial pollution management and the structure of political institutions may be very different in Malaysia and Thailand. Identifying these differences and how they have affected pollution control policies requires closer examination of each country. We focus our attention first on Thailand and then on Malaysia.

Thailand

Despite the establishment of parliamentary democracy in 1932, domestic politics in Thailand has, until recently, been the sole purview of a small administrative elite within the Thai state. (Much of what follows, unless noted, is taken from Rock 1994) The failure of democratic institutions to take hold earlier was the consequence of an elite-based transition to democracy that did not depend on either popular-sector groups or an autonomous bourgeoisie. This absence ultimately led to strong bureaucratic control over business groups, trade unions, and farm organizations. This was complemented, after 1960, by bureaucratic and military elites ceding important roles to newly trained technocrats in peak institutions. As a result, until the 1990s, democratic political institutions remained weak and unstable. Political scientists dubbed this particular political formation a "bureaucratic polity" (Riggs 1966)—a polity where

> political parties are weak and incoherent, national parliaments subject to control by the executive branch, interest and client groups non-existent and military intervention in domestic politics constant, the only group which contests ... is the bureaucracy. (Chai-anan 1971, 9–10)

Dynamism in Thailand's bureaucratic polity was provided by an incessant rivalry among "big men" within the political elite over the division of spoils extracted from Sino–Thai "pariah" entrepreneurs. When one big man met another, conflict ensued. After one rival defeated another, a cycle of dominance followed, only to be repeated when the latest big man fell.

Christensen and others (1993, 21–24, 26–28) describe how divisions of spoils worked in this nominally democratic, but bureaucratic polity. Concurrent with the decision of Field Marshall Sarit Thanarat in the mid-1950s to rely on the private sector as the engine of growth, the government created the institutions for modern macroeconomic management and vested them with the authority to maintain stability (Muscat 1994, 92). Both legal rules and informal norms for budget preparation insulated the budget from bureaucratic politics.

Because there were so few opportunities for rent-seeking in budget policy and practices, rent-seekers turned to controlling the sectoral agencies—the ministries of industry, commerce, and agriculture and the Board of Investment. Rent collection within the sectoral agencies was facilitated by a separation of the macroeconomic agencies from microeconomic policymaking and by fragmentation of industrial policy across a wide array of sectoral agencies. Thus, after an election, the victorious party coalitions and their big men were routinely allocated ministries along party lines, and cabinet ministers used their portfolios to reward their business supporters. By itself, this combination of nondemocratic politics, weak interest groups (particularly popular groups), and a rent-seeking feudalization of the sectoral agencies by big men in the bureaucratic polity was sufficient to forestall the development of effective public-sector industrial pollution management programs.

Thus, it is not surprising that Thailand's first explicit environmental legislation, NEQA 1975, was not passed until popular groups were able to overthrow a military government in 1973.[13] Unfortunately, Thailand's popular democracy did not last. Following a bloody coup in October 1976, Thailand's most popular and democratic government was replaced by yet another military government. But the bureaucratic polity would never be the same. Significant changes in Thai society, the Thai state, and state–society relationships following a prolonged period of rapid economic growth shifted the balance of Thai politics away from rivalry between big men and toward liberal corporatism.[14]

Much of this shift was stimulated by the emergence of a substantial urban middle class, a stronger and more dynamic civil society, and an independent bourgeoisie (Girling 1981; Dalpino 1991, 64; Anek 1988). The growing financial independence of the Sino–Thai business community, especially the large conglomerates, more and more insulated business from the reaches of government (Suthy 1982, 2–27). Over time, government control of trade associations declined, and the business community began to play a more prominent role in Thai cabinets (*Bangkok Post* 1982, 31–32; Dalpino 1991, 64–65). These changes facilitated the emergence of stable semidemocratic rule in 1978 (Chai-anan 1990, 281–282), and the bureaucratic polity gave way to a broker polity, in which

> the key figure (became) the prime minister who (had) the main responsibility for brokering a free for all between a growing number of organised (particularly business) constituencies. (Ramsay 1985, 8)

The breakdown of the bureaucratic polity occurred alongside significant technical strengthening of the infrastructure of the Thai state. By 1986, nearly 40% of the top 400 civil servants in the country had master's or doctoral degrees from Western universities (Rock 1995, 753). This included significant numbers of senior officials in the so-called sectoral agencies. Although this was no guarantee of more developmentally oriented microeconomic decisionmaking, it provided an opportunity to rationalize industrial policymaking and to launch an export campaign. Prime Minister Prem Tinsulanonda, who governed from 1980 to 1988, seized this opportunity by relying on an economic crisis to significantly restructure the relationship between the core macroeconomic agencies and the sectoral ministries. A Council of Economic Ministers, a subcommittee of the cabinet, stood at the apex of this new relationship. The National Economic and Social Development Board (NESDB) acted as the secretariat for the Council of Economic Ministers (Muscat 1994, 178).

In 1981, Prime Minister Prem created a high-level forum, the Joint Public and Private Sector Consultative Committee (JPPCC). The JPPCC provided regular opportunities for dialogue between leaders of the business community and senior government officials in core macroeconomic agencies and sectoral ministries. In 1983, Prem established a subcommittee of the JPPCC chaired by the NESDB that included senior officials of sectoral agencies and business community leaders. The subcommittee set the agenda for monthly JPPCC meetings and tracked the implementation of JPPCC actions. This was followed by the creation of government–business industry-specific councils and government–business sectoral ministry councils. Thus, the Thai government moved cautiously toward a relationship of embedded autonomy with business, particularly big business.

These institutional changes within the government and between the government and the private sector provided a unique opportunity to successfully reform industrial policies and harness microeconomic policy changes in the sectoral agencies to the promotion of manufactured exports. As we now know, that effort was very successful (Rock 2000b). But rationalization of industrial policy and the microeconomic (sectoral) agencies during the era of liberal corporatism never addressed environmental issues. This, no doubt, reflected the probusiness, progrowth approach of the Prem governments, which for all intents and purposes excluded popular-sector participation in policymaking. Thus, Thailand's technocratic broker polity was no more interested or successful in addressing industrial pollution than had been its predecessor, the rent-seeking bureaucratic polity.

If the liberal corporatism of the Prem period had been sustained by subsequent democratic governments, it is just possible that Thailand might have been more successful at reducing industrial pollution. This possibility is best illustrated by the environmental policies of the highly technocratic and appointed, but short-lived, Anand Panyarachun governments of the early 1990s. Those governments pushed through NEQA 1992, which created the PCD of MOSTE, enhanced the role of NGOs by granting them legal status in the courts, regularized the EIA process, and institutionalized a citizen's right to know. Alongside this, the NESDB's Seventh Plan (1992–1996) established environmental targets for the first time in Thai history, and the NESDB pressed for the implementation of these targets by integrating environmental criteria into the Board of Investment's promotional criteria. All in all, these changes were impressive and continue to mark the high point in government concern and action regarding industrial pollution.

Unfortunately, the more democratic governments that followed the Anand governments returned to the microeconomic (sectoral) rent-seeking policies of the bureaucratic polity. As in the past, political parties continued to lack grassroots bases or coherent programs (Christensen 1991, 100). More often than not, they continued to be little more than loose affiliations of big men interested in capturing enough seats in parliament to win one or more cabinet positions (Callahan and McCargo 1996, 378). As in the past, the cabinet positions then were used to dispense favors, reward supporters, and build war chests for future elections.

For the first time in modern Thai history, however, elected politicians began to politicize macroeconomic policymaking. In 1988, the Chatichai Choonhavan government bypassed the NESDB and funded a number of pork-barrel projects that allocated lucrative public-sector contracts and trade quotas to loyal political supporters (Christensen et al. 1993, 100). Following the election success of the Chart Thai Party in 1995, a quintessential rural-machine patronage politician, Barnharn Silparcha, was named prime minister (King 1996, 136). His cabinet appointments and those for major economic ministries, particularly the Ministry of Finance, signaled the further politicization of macroeconomic policy (Murray 1996, 372). This was worsened by a corruption scandal that subsequently plagued the central bank, the Ministry of Finance, and the stock exchange. This gave the prime minister an opening to fire the heads of the Bank of Thailand, the Ministry of Finance, and the Securities and Exchange Commission. As a result, during his tenure Thailand saw the unprecedented coming and going of three separate ministers of finance.

This frontal assault on technocratic elements within the Thai state was being driven by several new rhythms in Thai electoral politics (Punyaratabandhu 1998, 164). To begin with, the consolidation of Thai democracy fostered a proliferation, rather than a reduction, of the number of political parties. As a consequence, no single party has been able to amass enough members of parliament (MPs) to sustain a majority in parliament. The result has been all too predictable: Fragile multiparty coalition governments have been formed on the basis of slim parliamentary majorities.

Under these circumstances, there has been little or no unity in government and almost no cohesion around policy issues, including environmental policy. As Hicken (1998, 1999) and MacIntyre (2001, 86) have argued, this particular institutional structure favors governmental delay of important policy decisions (an inability to enact and implement new policy initiatives or to commit to and sustain new policy initiatives) and the underprovision of public goods (e.g., environmental protection). As they also say, the structure favors the use of veto authority by minority parties to win support for pork-barrel projects. To make matters worse, the daunting task of running party candidates in multiple-constituency rural districts led candidates and parties to resort to rural vote buying to ensure the election of MPs (Ockey 1994). Parties and candidates turned to contributions from the Sino–Thai conglomerates in Bangkok and provincial businessmen to finance their vote-buying machines in the countryside (Christensen 1991, 101). When combined with the institution of a quota system, which allots cabinet positions on the basis of the number of MPs a party elects, money for rural vote buying became critical to electoral success. As was mentioned above, once elected, MPs and cabinet members used their political positions to dispense favors, reward their supporters, repay their debts, and build war chests for the next election.

Because these tendencies have gone largely unchecked in Thailand's new semidemocracy, rent-seeking by new political elites and their business cronies has seriously undermined macroeconomic policies, as well as the microeconomic policies of the technocratic broker polity of the Prem period. Both the public and the environment have paid a high price for this particular variant of elite-based democracy with its attendant institutional structure. Capture of the state by business and an institutional structure that rewards delaying important decisions, the underprovision of public goods, and the use of veto power to attain political pork has meant that the demands of the public go largely unheard by politicians. This, no doubt, explains why successive democratic governments have failed to respond to pollution incidents, such as those around EGAT's 13 power plants at Mae Moh and at IEAT's industrial estate at Map Ta Phut.

It is clear that the Thai government's elite-based business politics—in the old bureaucratic polity, the more recent technocratic broker polity, and the new semidemocracy—have been sufficient to forestall better industrial pollution management. In the old bureaucratic polity, concern for the environment took a back seat to the rent-seeking cabinet members who treated the microeconomic agencies as their own personal fiefdoms. Although rent-seeking by cabinet members during Thailand's technocratic broker polity gave way to a rationalization of industrial policies that facilitated an export boom, the broker polity's focus on rationalizing sectoral policies and cultivating government–business ties to facilitate economic growth meant that little attention was paid to worsening industrial pollution. Unfortunately, the return to and consolidation of democracy after the end of the broker polity led to the capture of the state by business interests and an institutional structure that makes it difficult for governments to enact and implement new policies. When combined with vote buying to win elections, political parties that lack ideas or a mass base, and weak and fragile coalition governments, the result was predictable: There was a return to the rent-seeking sectoral policies of the bureaucratic polity and politicization of macroeconomic policy. Under these circumstances, it is not surprising that concern for the environment fell by the wayside.

Malaysia

Politics in independent, semidemocratic Malaysia has been profoundly affected by the past: British colonial rule relied on an ethnic division of labor to extract a surplus from an export enclave economy based on tin and rubber.[15] In this system, indigenous Malay peasants grew rice and cultivated rubber on small plots, while elite Malays served as hereditary aristocrats and public servants in the prestigious Malay Administrative Service (Crouch 1996, 15–16, 18). Poor Chinese immigrants worked in the tin mines, while (as in Indonesia under Dutch rule) their well-to-do brethren served as traders, businessmen, and commercial intermediaries between the British trading houses and the indigenous Malay population (Bowie and Unger 1997, 70). Moreover, poor immigrant Tamils from India worked on British-owned rubber plantations, while the more educated Tamils moved into the professions and became shopkeepers. By the late 1920s, more than half of the 4 million people in British Malaya consisted of Chinese and Indian immigrants (Snodgrass 1980, 18).

British Malaya remained a multiethnic colony with a multiethnic division of labor until World War II. After the war, when the British returned to Malaya, the colonial government attempted to rationalize its administration; as part of that effort, the government proposed eliminating the special privileges enjoyed by indigenous Malays (Case 1995, 79). Indigenous Malay opposition to this proposal was swift and vociferous, and it cut across class lines (Crouch 1996, 36). In 1946, members of the Malay Administrative Service took advantage of the opportunity created by this crisis to form an organization, the United Malay National Organization (UMNO) (Crouch 1996, 17).

A few years later, UMNO was the only major Malay organization in British Malaya. Subsequently, the British indicated a willingness to grant independence, assuming a way could be found to ensure domestic tranquillity in multiethnic Malaya. Officials of UMNO responded to this concern by joining with a conservative Chinese organization, the Malayan Chinese Association (MCA), in a political alliance against a Malay-led multiethnic party for the 1952 elections in Kuala Lumpur (Crouch 1996, 19). The alliance between UMNO and the MCA proved to be a great electoral success; subsequently, an Indian organization, the Malayan Indian Congress (MIC), joined them. Together, UMNO, MCA, and MIC—known as the Alliance—won every national election between 1955 and 1969.[16]

At Malaysia's independence in 1957, the UMNO-dominated Alliance government inherited a classic export enclave economy with a multiethnic division of labor (Salleh and Meyanathan 1993, 21–22). Primary commodity exports equaled 55% of gross domestic product, and manufactures were less than 10% of GDP. Forty-nine percent of the population was Malay, 37% was Chinese, and 12% was Indian. To make matters worse, the indigenous Malays were overwhelmingly poor rural dwellers who survived as rice farmers and smallholder rubber growers. Their Indian and Chinese counterparts, conversely, were much less likely to live in rural areas or to be poor (Bruton et al. 1992, 201).

The government of newly independent Malaysia—thus based on a multiethnic political coalition, and facing enormous disparities in income and well-being between the indigenous Malays and Chinese and Indian immigrants—decided to implement a three-pronged development strategy. Because senior officials in government were imbued with a conservative fiscal tradition (which emphasized the importance of balanced budgets, low inflation, a competitive exchange rate, and the balance of payments), the first prong was that the government maintained conservative macroeconomic policies (Salleh and Meyanathan 1993, 21). Although the "bargain" that had created the Alliance granted the Malays and UMNO dominance in politics and the civil service, it also protected Chinese business interests (Case 1995, 79). This combination put the UMNO-led government in a difficult position. It wanted to promote a more industrialized and diversified economy, but it could not be seen as actively promoting Chinese industrial interests.

The second prong of the strategy, therefore, was to resolve this contradiction by adopting a laissez-faire approach to industrial development complemented by mild import-substitution industrial policies and a Pioneer Industries Ordinance that offered promotional privileges and infrastructure to foreign investors (Salleh and Meyanathan 1993, 4). Thus industrialization in independent Malaysia, as in Singapore, was driven by promotional privileges offered by government agencies to attract foreign multinational corporations. As might be expected with an UMNO-led government, the third prong of the strategy was that the government intervened heavily in the rural economy to help indigenous Malay paddy farmers and smallholder rubber growers. This intervention focused on modernizing rice agriculture and rubber production; diversifying the agricultural economy away from rubber, whose price was falling; and developing land for new crops, particularly palm oil. It is not surprising that nearly half of all development expenditures between 1956 and 1965 went into rural development (Bruton et al. 1992, 235).

Between 1957 and 1970, real GDP grew a robust 6.5% a year, real per capita income increased 25%, the share of manufactures increased from 10%

to 16% of GDP, and inflation was low (Bruton et al. 1992, 233; Crouch 1996, 21). As Bruton and others noted, the decade looked like a textbook case of growth with stability. But all was not well. Unemployment remained high, employment growth in manufactures was low, income inequality increased, and the incidence of poverty (particularly among indigenous Malays) remained stubbornly high—89% of those classified as poor lived in rural areas, and 74% of these rural dwellers were Malays. To make matters worse, the incidence of poverty among indigenous Malays was 65%, relative to 39% for Indian Malaysians and 26% for Chinese Malaysians. In short, Malaysia's conservative macroeconomic policies and its focus on increasing growth in industry and agriculture had not trickled down to rural Malays.

This combination fed growing frustrations within the Malay community, and it led to calls for the government to do more to improve the economic position of indigenous Malays. At about the same time, ethnic tensions between the Malays and Chinese began to worsen following Singapore's entry into a federation with Malaysia. When the Singapore-led People's Action Party began playing an active role in Malaysian politics, Malays began to see this as a challenge to their basic right to govern Malaysia. Communal race riots between 1964 and 1967 led to the deaths of 53 people (Crouch 1996, 23). In this volatile atmosphere, the government was forced to deal with another contentious issue: whether English or Malay would be the national language. Although the government reached a compromise, making Malay the national language and English the official language, many Malays considered this a betrayal.

All of these issues came together in the 1969 elections, during which non-Malay opposition parties challenged both the MCA and the MIC (Crouch 1996, 23, 25). Although the Alliance won the national election, UMNO lost in Penang and was seriously challenged in Perak and Selangor by an Islamic party, Parti Islam Se-Malaysia (PAS), whereas non-Malay opposition parties won 24 seats in the new parliament and MCA representation fell by 50%. A postelection celebration by non-Malays in Kuala Lumpur precipitated a riot on May 13. The riot shocked the Malay elite, who now recognized that the government would have to take a more active role in improving the lot of indigenous Malays. The result was promulgation of the New Economic Policy (NEP) in 1971. The NEP aimed to reduce the incidence of poverty from 50% in 1970 to 20% in 1990, achieve a distribution of employment across sectors and occupations that reflected the racial composition of Malaysia, and increase indigenous Malay ownership of share capital to 30% by 1990.

The achievement of these ambitious goals required a multifaceted program. Preference was given to Malays in secondary and university school admissions, and educational opportunities were expanded. Preferences in school admissions were reinforced by preferential hiring practices. To increase Malay ownership in share capital, the government used public funds to acquire shares in foreign, largely British, companies. Once acquired, those shares were deposited in Permodalan Nasional Berhad (PNB, the National Equity Corporation), which then sold its shares in PNB to indigenous Malays (Pletcher 1991, 630). This "velvet" nationalization of foreign ownership was designed to increase indigenous Malay ownership of share capital without endangering growth (Salleh and Meyanathan 1993, 5). Finally, the government invested heavily in rural development, agricultural modernization, and diversifying the agricultural economy.

The modernization of agriculture focused on rice, rubber, and palm oil. Rice, Malaysia's basic food crop, and rubber were grown by indigenous and poor Malay peasants on small plots. The earliest support for rice agriculture focused on drainage and irrigation projects, research on high-yielding varieties of rice that performed well under Malaysian conditions, subsidies for modern inputs, and a modest government-supported price for rice (Mehmet 1986, 37, 57). These interventions were designed to raise farmers' incomes, increase the double cropping of rice, and improve rice self-sufficiency. After the introduction of the NEP in 1971, the government created the National Padi Rice Authority (LPN—Lembaga Padi dan Beras Negara).

Over time, the LPN became a vehicle for heavily subsidizing rice prices (Tamin and Meyanathan 1988, 107, 130). This was achieved by stabilizing the price of rice around a level substantially higher than the world price of rice (Rock 2000d). Government support for rubber dates from 1952, when the colonial government established a Rubber Industry Planning Board (Salleh and Meyanathan 1993, 29). Despite government efforts to rejuvenate the industry, the incidence of poverty among smallholder rubber producers was 65% in 1970 (Bruton et al. 1992, 281). Following substantial government investment and support for smallholding rubber producers through the Rubber Industry Small-Holder Authority, the incidence of poverty among these producers had declined to 43% by the mid-1980s.

The modernization of palm oil was part and parcel of the government's effort to diversify the rural economy away from rubber, a commodity whose price was in long-term decline. Because rubber, like rice, was cultivated on small plots by poor indigenous Malay farmers, expansion into more profitable palm oil held out the prospect of increasing the incomes of poor Malay farmers and reducing the incidence of poverty. As was mentioned above, FELDA became the government's primary vehicle for expanding the smallholder oil palm sector. FELDA cleared new land; built the infrastructure, including housing for settlers and for palm oil production; and provided technical assistance to settlers who moved into FELDA's nuclear estates.

Although this program has been costly, the available evidence suggests that it has also been quite successful.[17] Along with the rice intensification and subsidy program and the rubber modernization program, it has also been an important political tool for UMNO in the Malay countryside. This was and is important to UMNO, because the 1957 constitution essentially guaranteed Malay dominance in politics by granting greater weight to rural election districts than to urban constituencies. Initially, rural districts could be no smaller than 85% of the size of the average constituency. In 1973, however, this limitation was abolished, so that by 1982 the largest constituency in peninsular Malaysia had more than four times the number of voters as the smallest rural constituency (Crouch 1996, 59). Given the greater share of indigenous Malays in rural areas, this gerrymandering of election districts virtually guarantees Malay control of parliament.[18] Because UMNO has been contesting with PAS and other opposition parties for control of the rural Malay vote, and hence the Malay state, its rural development programs have been an important source of patronage and electoral support (Scott 1985).

A range of evidence suggests that UMNO has effectively used its rural development programs to reward its supporters and punish its political enemies (Crouch 1996, 40–42). Local village heads and villagers who control village development and security committees tend to be UMNO stalwarts, as do those who evaluate applications by villagers for land titles. The distribution of subsidized agricultural inputs tends to be controlled by an UMNO representative in villages. It is not surprising that UMNO-affiliated farmers have been the primary beneficiaries of subsidized rural credit. There is evidence that settlers in FELDA's land development schemes were expected to support UMNO. In at least one instance, a minister of land and regional development openly complained that too few of FELDA's settlers were UMNO supporters. And there is evidence that local UMNO officials have used their positions to block agricultural and basic-amenity projects in areas controlled by opposition political parties (Shamsul 1983, 472).

The importance to UMNO of political support from the rural Malay peasant can also be seen in the government's reaction to protests by Malay farmers (Crouch 1996, 91). In 1974, after a large-scale demonstration in Kedah protesting a big decline in the price of rubber, the government introduced a rubber price-support scheme. In 1980, the government revised a newly announced coupon system to pay the rice subsidy after a demonstration against the system. In addition, it detained ordinary citizens and arrested seven PAS officials under the Internal Security Act. These actions suggest that UMNO-dominated governments fully recognize the need to sustain the support of rural Malays, particularly against the political threats posed to it.

How, then (if at all), did UMNO's concern for its rural base of support affect the government's industrial pollution management policies? Although there is virtually no research on this topic, it is possible to piece together a relationship between the two. As was mentioned above, palm oil and CPO production expanded rapidly after the NEP as the government sought to improve the lot of poor Malay peasants. Despite the fact that the settlers in FELDA's new palm oil schemes were not the poorest of the poor, the schemes were helping the poor and UMNO. But as CPO production expanded, CPO effluent became the major source of water pollution in many of Malaysia's rivers. By 1975, 42 rivers ceased supporting aquatic life or providing drinking water. As CPO effluent loads continued to rise, aerobic digestion of CPO waste turned anaerobic, giving off particularly foul odors.

Because of this situation, most of the public complaints about water pollution between 1974 and 1978 focused on CPO mills and their effluent (Lim 1977, 4–6, 39). Fishermen complained again and again that water pollution was killing fish populations and destroying their livelihoods. Rice farmers complained that the CPO waste dumped into the canals they used to irrigate their rice fields was lowering rice yields and killing fish (K.E. Lee 1978, 51). In another instance, the holding ponds for CPO waste overflowed, flooding villages and destroying homes and livestock (Rashid 1979). In a number of instances, local UMNO officials and local district officers asked the government to intervene to get CPO mills to clean up their pollution (Ramayah 1979, 14).

The UMNO-led government—faced with these complaints and fearing that its political rivals in the countryside might capitalize on indigenous rural Malay unrest—had no choice but to find a solution to growing CPO effluent. Because the government could not shut down the CPO mills without undermining the success of its palm oil schemes, the government launched an integrated attack to clean up CPO waste. Once an economically viable treatment technology had been discovered, DOE ratcheted up emissions standards, ultimately delinking CPO production from CPO effluent. Given the tight links among elites within the Malaysian state, it is inconceivable that Malaysia could have so successfully delinked CPO production from CPO waste without high-level support from UMNO.[19]

Conclusions

Malaysia and Thailand provide stark contrasts in the evolution of public policies designed to reduce industrial pollution. Although Malaysia has been quite successful in getting industrial facilities to reduce their emissions, Thailand has found it next to impossible to get industrial polluters to reduce their pollution. These different experiences provide sobering lessons on how domestic politics and the design of governmental institutions in democracies and during democratization affect the success of policies to manage industrial pollution.

On the surface, Malaysia's successful experience with industrial pollution management looks similar to that of South Korea and Taiwan, where democratic governments responded to growing public pressure from organized and unorganized groups in civil society to clean up the environment.[20] But closer examination of Malaysia's experience suggests that it is significantly different from those of South Korea and Taiwan. In South Korea and Taiwan, democratization of longstanding authoritarian regimes, rather than democracy per se, unleashed environmental protest movements that led newly elected democratic governments to create capable, more and more effective command-and-control regulatory agencies (Tang and Tang 1997; Aden et al. 1999; Eder 1996). Nothing like this ever happened in Malaysia.

To begin with, Malaysia has been semidemocratic since achieving independence (Case 1993, 185, 187). Elections are regularly contested, opposition parties are permitted, and independent groups in civil society are tolerated and permitted to raise mass grievances. But the government seriously constrains each of these. The potential electoral success of opposition political parties is hampered by malapportionment of electoral districts, bans on open-air rallies, and government control of the media. Civil society organizations in Malaysia are generally controlled by the state; and the state has not been averse to using its police powers, including an Internal Security Act, to detain representatives of popular groups and to undermine them (Crouch 1996, 29–30).

In addition, under the Societies Act of 1966, any organization of more than seven persons is required to officially register with the Ministry of Home Affairs (Barraclough 1984, 451, 454). Amendments to the Societies Act in 1981 extended government control over independent organizations in civil society by allowing the government to order societies to amend their constitutions and by denying societies the right to influence the policies or actions of either the federal or state governments. Government control of independent political activity extends to students, university professors, and union members, all of whom are barred from political activities. It also extends to the countryside and Malay villages, where UMNO officials control the resources allocated for rural development through a wide range of local committees that are responsible for, among other things, pollution (Shamsul 1983, 459).

Because of the government's constraints on independent activity by organized groups in civil society, environmental protest (including that in the countryside) has not been led by well-organized environmental NGOs, as in South Korea and Taiwan. In one of the few instances in which environmental NGOs have successfully organized environmental action—against the government's plans to build a hydropower plant in the Taman Negara National Park—the government reacted by arresting 10 of the NGOs' members under the Internal Security Act (McDowell 1989, 321). In these ways, Malaysia looks a lot like its rich neighbor Singapore to the south. But this does not mean that the government does not respond to public pressure. In fact, as the examples given above have shown, it does. This is because, as Case (1993, 187) says, the limited activities of opposition political parties and interest groups act as safety valves for societal discontent, while providing important feedback to UMNO and the government.

This is particularly true of protests and complaints from rural Malays. Because the government is contending for the support of rural Malays with PAS—the Malay and Islamic opposition party—both it and UMNO are extremely sensitive to rural unrest, particularly in northern peninsular Malaysia. Although most of the palm oil plantations and CPO mills are in southern Malaysia, some are in the north, and at least some of the complaints about CPO pollution emanate from there.[21] In addition, poor rice farmers in the north often hope to be chosen for migration to one of FELDA's palm oil schemes (Scott 1985, 55). Thus, failure to control pollution from mills located anywhere in the country could well undermine support from poor rural Malays who migrated to one of FELDA's palm oil schemes. Failure to reduce CPO pollution might also subject UMNO to a criticism from conservative Islamic forces that they are failing to protect indigenous Malays—who are, after all, "sons of the soil"—from exploitation by the Chinese and foreigners (McDowell 1989, 320–322). Given these fears, it is not particularly surprising that the government has acted to clean up CPO pollution in the country's rivers.

But this raises another issue: Why has the Malaysian government apparently been able to both design and implement new environmental policies and to sustain commitment to these policies over time, unlike successive democratic governments in Thailand? Part of the answer lies in the fact that governmental institutions in Malaysia, unlike those in Thailand, are overwhelmingly dominated by one party: UMNO. This means that the Malaysian government does not face the multiple veto threats that Thai governments regularly experience. It also means that Malaysia has a much greater ability than Thailand to enact and implement new policies (MacIntyre 2001, 88, 99–102). In MacIntyre's terminology, Malaysia has more than ample policy flexibility, because it is basically a unitary state with just one collective potential vetoer, its longstanding and oversized coalition: Barisan Nasional. Although this can contribute to policy instability, when vital interests are threatened—as they were with growing disenchantment among rural Malays over CPO pollution—the government has acted decisively and resolutely.

In the Thai case, successive governments—in Thailand's bureaucratic polity, its technocratic broker polity, and in its semidemocracy—have been unable or unwilling to grant regulatory agencies either the resources or legal authority to monitor and enforce tough air and water emissions standards. Moreover, efforts to lean on leapfrogging environmental strategies (e.g., a community's right to know, public pressure, and external pressure) have not worked very well. The question is, Why? Part of the answer surely lies in the fact that the success of the strategies tried by Thailand depends on a regulatory base, which must be provided by a competent, pragmatic, honest regulatory agency. Because this agency does not exist, it is not particularly surprising that Thailand's strategies have not worked.

This situation, however, raises another question: Why has democratic Thailand been unable to create and empower such an agency? One might not have expected much environmental action from Thailand's bureaucratic polity or its technocratic broker polity, but experiences elsewhere in East Asia suggest that democratization should have been a powerful force for environmental change.[22] This has not happened in Thailand for several reasons. To begin with, Thai democratization was elite based rather than mass based. This had two effects. First, it limited the influence of popular pressure on government. Second, it expanded the influence of the business community, particularly large firms in Bangkok and the provinces. Businesspeople, particularly in the provinces, have been elected to parliament in increasing numbers, cabinet members come more and more from the business community, and businesspeople finance the vote-buying machines in the Thai countryside that are so critical to winning elections and cabinet portfolios.

Unfortunately, the development of civil society groups in Thailand, particularly environmental NGOs, has not kept pace with the penetration of politics and government by business. But this is not to say that environmental NGOs do not exist in Thailand—they do. Rather, the problem is that most focus on natural resource issues rather than on industrial pollution.[23] Most are poorly financed and depend on external support (Komin 1993). Few have a long-term, professional staff. Most are poorly networked and spread across a large array of environmental issues. And most have little access to the media (So and Lee 1999, 128–130, 131–137). Of those environmental NGOs that focus on industrial pollution, virtually all are heavily supported by the business community—this includes the Thailand Environment Institute, the Thailand Environment and Development Network, Magic Eyes, and the Thailand Business Council for Sustainable Development.

To make matters worse, the structure of governmental institutions in Thailand—the electoral system, the legislature, and parties—rewards, as Hicken (1998, 1999) argues, policy delay and the underprovision of such public goods as environmental protection. They also encourage, as MacIntyre (2001) argues, the use of veto power to thwart most policy initiatives, particularly those such as environmental protection that threaten business interests. This particular combination—business-dominated democratic governments, weak environmental NGOs, and an institutional structure that works against new policy initiatives—does not augur well for pollution reduction in Thailand.

Several clear lessons emerge from the Thai experience. To begin with, the Thai case provides a concrete example of how and why developing-country democracies might not do well at protecting the environment. This is sobering to those who pin hopes for environmental improvement on democratization or who argue that it is a necessary condition for improvement. Beyond that, the Thai case provides insight into what makes democracies—and what should make Thai democracy—more responsive to environmental concerns. The nature of the relationship between the public and its democratic government is critical to successfully meeting environmental challenges. Governments that are unresponsive to popular pressure, such as Thailand's, are not likely to be very successful unless, as is Singapore's, they are led by near-benevolent despots with environmental concerns. It is interesting that a situation similar to that of Thailand has developed in Chile, where an explicit bargain between the ruling Christian Democrats and the business and military communities has blocked the creation of a separate ministry of the environment to monitor and enforce tough emissions standards (Silva 1997, 23).

How, then, does one make a democratic government like Thailand's more responsive to popular pressure? More openness, transparency, and accountability in government no doubt would help. But even if the Thai government becomes more responsive to public pressure, unless its institutions evolve and thus grow strong enough to enact, implement, and sustain commitment to new environmental protection policies, there may not be much of an opportunity to improve ambient environmental quality. Some of this evolution, however, might be accomplished through the constitutional and electoral reform Thailand is now pursuing (Klein 1998).

Finally, even if the Thai government's institutions evolve toward greater effectiveness, its industrial pollution management is not likely to improve unless it creates a strong command-and-control environmental regulatory agency. That is what was required to achieve lasting environmental improvement in the countries that belong to the Organisation for Economic Co-operation and Development and in those East Asian economies that have successfully tackled their pollution problems. Given this, the failure of the Thai government to create such an agency makes its focus on a community's right to know and external market pressure look like little more than a politically expedient diversion.[24]

Notes

[1]For a discussion of rising industrial pollution loads in Malaysia, see World Bank (1993, 44–50). For a discussion of rising industrial pollution loads in Thailand, see World Bank (1994b, 134–139).

[2]As of 1993, there were 43 sector-specific acts with environmental provisions (Sani 1993, 70).

[3]The concentration of total suspended particulates (TSP) in the air in Bangkok was 370 micrograms per cubic meter (μg/m³) of air in 1998 (World Bank 2000b, 5). Of four rivers north and south of Bangkok, two had dissolved oxygen levels below the Thai standard of 4 micrograms per liter (μg/l), and three out of four failed to meet the Thai biological oxygen demand (BOD) standard (World Bank 2000b, 6).

[4]By 1998, average air quality throughout the country was rated good (below 90 μg/m³ of air) (DOE 1998, 1).

[5]Because the Pollution Control Department (PCD) of the Ministry of Science, Technology, and the Environment (MOSTE) has not set emissions standards for EGAT's power-generating plants, EGAT has set its own standards, and it monitors and enforces them. At Mae Moh, where EGAT has 13 power-generating plants, it has 16 monitoring stations that monitor for sulfur dioxide, nitrogen dioxide, and TSP (World Bank 1994b, 74).

[6]There were 126,000 industrial plants in Thailand in 1996 (Sunee and Canino 1998).

[7]Participation in the pilot project is limited to companies that are certified according to the nonenvironmental quality standard known as ISO 9000, and it is conditioned on support from a company's top management and assignment of two persons within the company who will work on this project full time (interviews in Thailand, 1996).

[8]Thailand's PCD does not even have the authority to monitor and inspect factories. This responsibility lies with the DIW in the MOI.

[9]The DOE's water quality index takes account of five water pollutants—biological oxygen demand, ammoniacal nitrogen, suspended solids, pH, and chemical oxygen demand (Sani 1993, 38).

[10]With the enactment of a new constitution in Thailand in 1997, the World Bank began working with MOSTE to strengthen its environmental institutions. Of course, several bilateral aid agencies are working in both countries, but mainly in the area of pollution prevention. Denmark's aid agency has a prevention project in both countries. The United States has also been working on clean production, as have the Japanese in Thailand. But, for the most part, these programs are peripheral to the environmental challenges both countries face. This is particularly so in Thailand.

[11]For arguments about why democratic governments may do better at industrial pollution management than authoritarian governments, see Tang and Tang (1997), Payne (1995), and Congleton (1992).

[12]There have been repeated pollution incidents at Mae Moh, where the EGAT has 13 power plants, and at an industrial estate at Map Ta Phut on Thailand's Eastern Seaboard. For a discussion of the incidents at Mae Moh, see Somsak (1997, 1–4; 1998, 1–2; 1999, 1–2), and Wasant (1999, 1–3). For a discussion of the incidents at Map Ta Phut, see Limanon (1999, 57–60).

[13]For a discussion of Thai democracy between 1973 and 1976, see Morell and Chai-anan (1981).

[14]For an extensive discussion of this shift, see Anek (1992).

[15]For a discussion of the colonial period in Malaysia, see Jomo K.S. (1986). For a discussion of Malaysia's semidemocracy, see Case (1993).

[16]The Alliance was renamed the Barisan National Front after it subsequently expanded to include a number of other opposition parties (Crouch 1996, 31).

[17]Bruton et al. (1992, 235) state that most analysts find an internal rate of return for FELDA land development projects to be above 10%.

[18]As Crouch (1996, 59) says, Malays formed a majority in 57% of peninsular Malaysia's constituencies in 1964, whereas they formed a majority in 69% of constituencies in 1978 and 70% by 1986.

[19]As Crouch (1996, 19–20) says, Malaysia's political system is dominated by a relatively homogeneous elite that shares similar values, orientations, and education. Because of the cooperation between UMNO, MCA, and MIC, the political system has some of the characteristics of consociationalism, which achieves political stability through the elite-level integration of plural groups. As he also says (p. 33), those who manage politics work to settle issues behind closed doors.

[20]For a discussion of Taiwan, see Chapter 3. For a discussion of South Korea, see Eder (1996) and Aden et al. (1999).

[21]In 1977, farmers in several areas in Kedah complained about lower rice yields and the death of fish in canals and rice fields brought about by CPO pollution. At about the same time, the Environmental Protection Society of Malaysia called on DOE to immediately control CPO effluent discharges from mills in Trengganu (K.E. Lee 1978, 51).

[22]This has happened in Taiwan (see Chapter 3) and South Korea (Eder 1996; Aden et al. 1999).

[23]This becomes clear from reading the literature on environmental NGOs. See, for example, So and Lee (1999), Hirsch and Lohmann (1989), Komin (1989), Lohmann (1991), and Quigley (1996).

[24]This also holds for the use of market-based instruments, about which there has been a recurrent discussion in Thailand; see, for example, Suphavit (1996). To date, however, these discussions have not had much effect.

7

Comparing NIEs' Experiences with Controlling Pollution

As the case studies have demonstrated, there is substantial variation in the degree to which governments in the East Asian newly industrializing economies (NIEs) are successfully addressing the environmental problems attending high-speed urban industrial growth (see Figures 7-1 and 7-2 and Table 1-2). Singapore has achieved and sustained a high level of ambient air quality, as determined by the Organisation for Economic Co-operation and Development (OECD), and it has a low organic water pollution intensity of industrial output. Malaysia and Taiwan are approaching OECD levels of ambient air quality, and they also have relatively low organic water pollution intensities of industrial output. China has been able to improve ambient air quality in a significant number of the country's major cities, but these improvements fall significantly short of OECD levels. At the same time, ambient environmental air quality in many of the country's smaller cities continues to deteriorate, and the organic water pollution intensity of industrial output is high. Even though Indonesia has been able to get large industrial water polluters to reduce their wastewater emissions, the organic water pollution intensity of industry is relatively high, and this effort to reduce industrial wastewater emission has had little effect on ambient water quality. Moreover, Indonesia has not made much progress in reducing industrial air emissions. When combined with growing vehicular emissions, ambient air quality in major cities, such as Jakarta, remains poor. And to date, Thailand

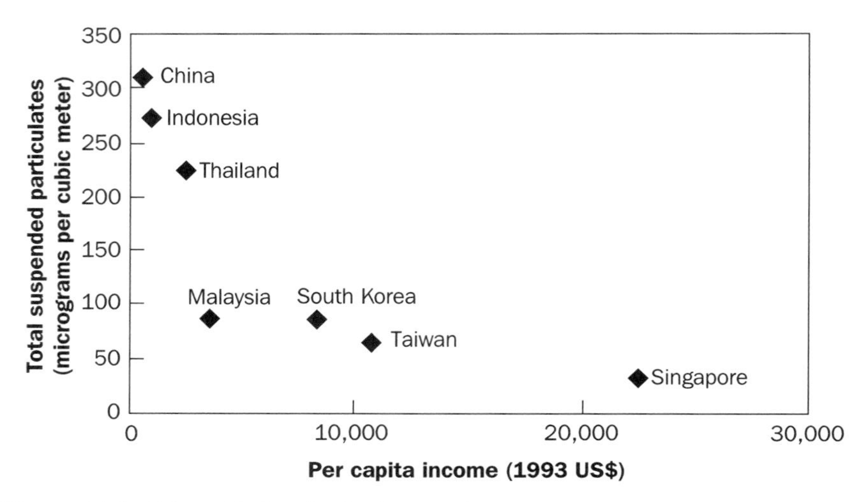

Figure 7-1. Urban Air Quality in Capital Cities in East Asian Newly Industrializing Economies

Sources: World Bank (2000a, 163); "Comparison of Air Quality with Other Countries," July 7, 2000, from the Web page, http://www.epa.gov.tw/english/offices/f/bluesky/bluesky3.htm.

has had the least success in reducing industrial pollution and improving ambient environmental quality.

Lessons from the Case Studies

Why are these environmental outcomes so different? The case studies demonstrate that they reflect differences in pressure to clean up; in type of political regime (authoritarian or democratic); in the ability of the government to enact, implement, and sustain new environmental policies; and in the extent to which environmental ideas have been borrowed from the OECD. That is, they depend on the very conditions that were identified in Chapter 1 as affecting the choice of strategies to control industrial pollution and the evolution of those strategies.

Singapore, Malaysia, and Taiwan

In Singapore, the commitment of a near-benevolent despot and a strong, autonomous state with substantial embedded autonomy enabled the government to implement a top-down approach to pollution management that

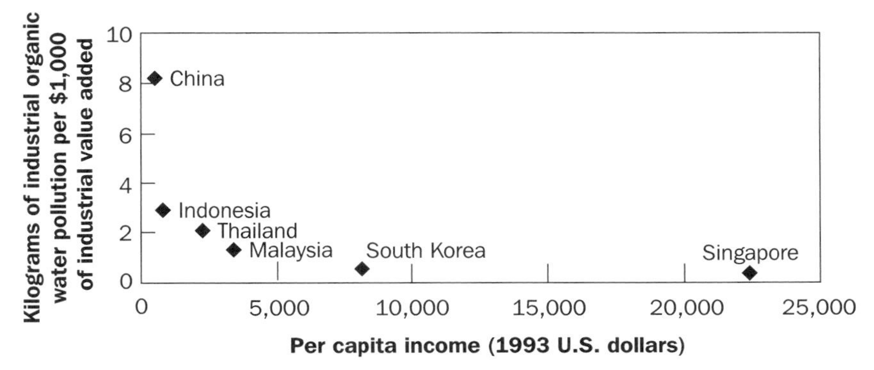

Figure 7-2. Water Pollution Intensity of Industrial Value Added in East Asian Newly Industrializing Economies

Note: Data for Taiwan were not available.

Sources: World Bank (1998, 236–237; 2000a, 134–136).

involved integrating a new, but very traditional, command-and-control environmental agency intc existing institutions of industrial policy. This meant that regulatory policies (or what Evans refers to as "policing policies"; 1995, 13) were added to the government's longstanding promotional policies, thus enabling the government to communicate to industrial polluters that it was serious about cleaning up pollution. It also enabled the new regulatory agency to take advantage of the close links extant between the industrial policy agencies and the business community. As Chapter 2 demonstrated, these close links were critical to Singapore's pollution management success. This all happened very early in the country's industrial transformation. Because of this, Singapore was able to grow rapidly without significant long-run deterioration in ambient environmental quality—in other words, it was able to grow while cleaning up the environment.

Semidemocratic Malaysia, like Singapore, also found a way to clean up while growing rapidly. But unlike Singapore, the pressure to clean up came from outside government and from an important voting constituency: rural Malays who expressed increasing dissatisfaction over rising pollution loads from the numerous crude palm oil (CPO) processing mills that dotted the Malay countryside. Because the dominant party in government, the United Malay National Organization, depended on votes from rural Malays for its control of government, it could not easily ignore this constituency or the inroads made by an important Islamic opposition party. This combination enabled and forced the government to draw on its considerable state strength and embedded autonomy, as in Singapore, to create and empower a very traditional command-and-control environmental agency to clean up CPO pollution. But, as in Singapore, it did more than just impose tough technology-based emissions standards on CPO polluters. It used its embedded autonomy with a quintessential industrial promotion agency, the Palm Oil Research Institute of Malaysia (PORIM), to time its ratcheting up of emissions standards to the development within PORIM of the best available control technologies not entailing excessive cost. This made it possible to almost totally delink CPO production and exports from CPO emissions.

Taiwan—another government with a strong, autonomous state, substantial embedded autonomy, and capable institutions of industrial policy—unlike Malaysia and Singapore, pursued a "grow first, clean up later" environmental strategy and only seriously turned attention to pollution control after democratization. Earlier environmental neglect was the consequence of the tight link between government and business that focused on growth and exports at the expense of nearly everything else. But democratization spawned a large, vociferous environmental protest movement, including opposition political parties, that the government could not ignore. It is surprising that this previously authoritarian government won the country's first democratic election and built new relationships with old constituencies (academics) and new ones (environmental nongovernmental organizations, or NGOs) on a new issue (environmental cleanup) at the same time that it created a traditional command-and-control agency. This required the government, as in Malaysia and Singapore, to incorporate policing policies into what had been a largely promotional relationship between government and business (Wade 1990)—suggesting that even old, authoritarian governments and parties can be engaged in environmental cleanup, if their vital interests are threatened.

But unlike Malaysia and Singapore, the government of Taiwan's initial environmental strategy bypassed its premier institutions of industrial policy because of opposition within those institutions to cleaning up. This put the brunt of the cleanup on a new command-and-control agency and domestic public pressure. Once it became clear that this strategy was working, the industrial policy agencies responded by developing their own unique environmental improvement programs. The net result, as in Malaysia and Singapore, has been substantial improvement in ambient environmental quality and the integration of environmental considerations into industrial policymaking, although in very different ways from Malaysia and Singapore. In Malaysia and Singapore, regulatory agencies used their embedded autonomy to find end-of-pipe solutions to pollution abatement that did not undermine either corporate profits or exports. In Taiwan, the government's embedded autonomy was used for environmental midwifery,[1] which included technical assistance for pollution prevention, informational assistance for exchange of wastes, promotional privileges for the creation of an export-oriented environmental goods and services industry, and a unique government-led search for cleaner technologies.

Despite differences in initial conditions, in levels of development, and in the nature of politics and political institutions, the governments of these three economies took advantage of their strong autonomous states with substantial embedded autonomy with the private sector. The former enhanced the governments' policy flexibility, making it possible for them to enact and implement new policies. The latter made it possible to draw on the trust gained by years of positive collaboration between government and the private sector so that reductions in emissions did not threaten profitability or exports. This combination made it easier to sustain commitment to new industrial environmental improvement policies. And it suggests, contrary to what some have argued, that these governments can be as successful with policing, or regulatory, policies as they have been with promotional policies.

It is worth repeating that policing policies have been very dependent on the creation of tough, competent, pragmatic, and fair command-and-control environmental agencies with sufficient capacity and legal authority to monitor and enforce new emissions standards. For the most part, these institutions were modeled on their counterparts in the OECD countries, particularly the U.S. Environmental Protection Agency (U.S. EPA). This suggests that there may be fewer political, social, and cultural impediments to such transfers than was previously thought. Where industrial pollution control strategies have differed in these economies is in how the new environmental agencies interact with both the public and premier industrial policy agencies.

China, Indonesia, and Thailand

What happens, however, when governments have less pragmatic, technocratic, and goal-directed bureaucracies? What happens if they also have less embedded autonomy with business? What happens if they are prone to more rent-seeking and corruption? And what happens if they try, as some have suggested (World Bank 2000a), to build their industrial pollution management strategies by taking advantage of policy innovations in the OECD countries and leapfrogging the old command-and-control model first tried in those countries? The case studies of China, Indonesia, and Thailand were designed to answer these questions.

Until recently, much of the environmental discussion in Thailand, among academics, within government, and in the NGO community, focused on leapfrogging strategies for environmental improvement—on external drivers, the most recent standards from the International Standards Organization (ISO; the standards are known as ISO 14000), eco-labeling, public disclosure policies, or the win–win opportunities created by pollution prevention. This is largely a consequence of the fact that successive democratically elected Thai governments have been unable or unwilling to create and empower a traditional command-and-control environmental agency, as in Malaysia, Singapore, and Taiwan. Unfortunately, there is virtually no evidence that any of these leapfrogging alternatives is working. This suggests that they may work, if they work at all, only after the creation of a successful command-and-control agency. That has been the pattern in the OECD countries, and it seems to be the case in Thailand.

Why have democratic governments in Thailand, however, unlike their counterparts in Malaysia, South Korea, and Taiwan, been unable or unwilling to create a strong command-and-control agency? It is not because of the absence of either a vibrant environmental movement or a free press that fails to report environmental accidents. Thailand has both, even though its environmental movement may be too focused on green issues and, at times, a little too beholden to the business community. It turns out that the problems in the Thai case lay in the nature of the country's democratic transition and in the structure of its democratic institutions. Because Thailand's democratic transition was elite based rather than broadly based (unlike that of Taiwan), the patron–client ties of the country's old bureaucratic polity were carried over into democracy. This means that the country's democratic institutions—particularly its political parties, legislature, and cabinets—have been essentially captured by business interests. As a consequence, individuals and popular organizations in civil society—unions, women's groups, and environmental organizations—have little influence on party politics, on the legislative agenda, or on decisions made by the prime minister and cabinet ministers.

To make matters worse, the structure of Thailand's democratic institutions—particularly its electoral system and its party politics—encourages rampant vote buying in rural areas by those with business backing and weak fragile multiparty coalition governments. This combination increases the number of potential vetoers in government; that is, those individuals and collective actors who can and do obstruct new policy initiatives, including environmental initiatives that businesspeople oppose. The overwhelming lesson from the Thai case is that democratization is no guarantee of environmental improvement. Both the nature of the transition to democracy and the structure of democratic institutions matter.

What can be learned from the Chinese and Indonesian cases? Politics and governmental institutions in both are distinctively authoritarian, but without the strong autonomous states extant in Malaysia, Singapore, and Taiwan. Instead, authoritarianism is fraught with patron–client ties, penetration of the state by business interests, and substantial rent-seeking. This does not appear to be a particularly strong brew for successful pollution management. Yet there are some, albeit partial and incomplete, successes in both. To be sure, neither government has created a robust command-and-control environmental agency. But both governments have gone much, much further than democratic Thailand. They have been slowly upgrading the capabilities of their environmental agencies, and they have drawn heavily on donor assistance to do so. They have also enacted and implemented a number of quite innovative pollution control policies.

The Indonesian case study demonstrates that even a purportedly soft, rent-seeking, bifurcated state—a state that is seen as quite good at macroeconomic policy, but quite ineffective at microeconomic policies—can get some industrial polluters to abate pollution if it sees a deteriorating environment as threatening its vital interests. In this way, the New Order government was similar to governments in Malaysia and Taiwan; it was, until the recent financial crisis, quite good at using protest as a safety valve, in coopting opponents to the regime, and at acting to protect its vital interests. But getting at least some polluters to abate their pollution required overcoming what was viewed as a congenital inability of the New Order government to enact and implement coherent microeconomic policies—which are pollution control policies par excellence. This proved possible because a strengthened national environmental impact management agency depended very heavily on its ties to the president and links to the World Bank to enact and implement both a voluntary pollution reduction program and a public disclosure program. Continuing political support from the president for both was critical; without it, it is doubtful that any of the large plants involved in either the voluntary pollution reduction program or the public disclosure program would have volunteered or allowed their environmental performance to be publicly disclosed. Technical support from the World Bank was equally critical; it offered an important veneer of technical competency.

Senior managers of Indonesia's environmental impact management agency, BAPEDAL, also took great care in the design and implementation of both of the PROKASIH and PROPER programs. They timed actions to take advantage of highly publicized pollution incidents. They used their limited technical capabilities to focus on a particular pollution problem: wastewater emissions from large factories. By doing so, they conserved scarce regulatory resources and limited regulatory capabilities. They took advantage of the fact that some large factories were already abating pollution. This meant that they were able to reward good performers. They developed clear, simple, credible, honest, and transparent systems to implement these innovative programs and reached out to polluters, environmental NGOs, communities, and the media to demonstrate how the programs worked. This inspired confidence in what they were doing and suggests that pragmatic attention to the political consequences of the institutional design of pollution control policies may be critical to their success.

It turns out that this pattern is consistent with other examples of the New Order government's ability to act decisively and pragmatically to make hard microeconomic policy choices. This suggests that the conventional view of the New Order state as soft, bifurcated, and fraught with rent-seeking patrimonial networks is at least somewhat overdrawn. But it needs to be noted that industrial environmental improvement in Indonesia stopped with these programs. This outcome reflected a lack of consensus within the New Order government over what to do about industrial pollution and the subsequent collapse of that government. And in the end, these programs did not contribute to improvements in ambient environmental quality, nor were they able to survive the turmoil surrounding the collapse of the New Order government and the instability that followed.

Something similar is at work in China, where a weak authoritarian government riddled with patron–client ties and substantial rent-seeking also invested heavily in its traditional command-and-control environmental agency, while allowing those in it to implement innovative pollution control programs. And as in Indonesia, creative managers in a strengthened national agency built and sustained important political support with their counterparts in cities, especially with mayors, for a program that annually rated, ranked, and publicly disclosed the environmental performance of the country's major cities. As in Indonesia, they kept this program relatively simple, credible, transparent, and honest. They also reached out to important constituencies, particularly in city-level industrial development and economic development agencies, to demonstrate how the program worked. This inspired confidence in what they were doing with constituencies whose support they ultimately needed. But, most important, they figured out how to integrate this program with the bargaining model of economic policy implementation that characterizes China. The net result has been some improvement in ambient environmental quality, at least in China's larger and more prosperous cities, and more than in either Indonesia or Thailand. But it remains to be seen whether this program can yield the kinds of improvements in ambient environmental quality that China needs.

The Chinese and Indonesian cases have several common elements that bear mentioning. They suggest that some minimum level of capacity in the environmental agency and some minimum amount of high-level political support for the agency may be necessary for such a nascent, weak agency to begin to make a difference. This helps to explain why so little has happened in Thailand, where the national environmental agency lacks both. It also helps explain why China has been able to achieve some improvement in ambient environmental quality, whereas Indonesia has not. It suggests that weak environmental agencies should consider starting small, innovating and focusing on particularly pressing problems. And it suggests the need to design innovations with care to overcome potential political objections. This combination helps to ensure that environmental improvement can begin to be seen well before the environmental agency acquires a full-blown capacity to mount a national monitoring and enforcement program.

General Lessons

Four other important general lessons emerge from the experiences of the East Asian NIEs. The first is the crucial role played by increasing openness to trade, investment, and new ideas. The second is the need to build command-and-control environmental agencies. The third is that embedded autonomy is needed to make these agencies effective. The fourth is that public pressure, in the context of these agencies' actions, can make a difference in the environmental policies and actions of all kinds of regimes. We examine each lesson in turn.

Openness Facilitates Environmental Improvement

Openness to trade and investment and to new ideas regarding the relationship between the environment and development have made it easier for those both inside and outside of government in the NIEs to pressure governments to take on the task of controlling industrial pollution. This increasing openness has meant that governments have had to prepare position papers for important international forums such as the United Nations Conference on the Human Environment in Stockholm in 1972 and the United Nations Conference on the Environment and Development in Rio de Janeiro in 1992. Openness thus has forced reluctant governments to directly confront interactions between development and the environment.

Openness also has exacerbated government officials' and industrial exporters' fears that the exporters might be shut out of important markets unless they improved their environmental behavior. This has led most of the NIEs' governments to launch ISO 14000 certification programs, and it has led private-sector actors to create business councils for sustainable development. It also has exposed government officials and the private sector to criticism by international NGOs that they were not doing enough to protect forests, biodiversity, and endangered species. And it has exposed governments to donor requirements to carry out environmental impact assessments of large infrastructure projects and to engage in building environmental regulatory agencies.

Openness to new ideas, particularly policy ideas, has meant that analysts both in and out of government have been able to influence policy by demonstrating the high costs of environmental degradation and the relatively low costs of cleaning up industrial pollution. This, no doubt, has made it easier to get several governments (Malaysia, Singapore, and Taiwan) to build and sustain successful command-and-control environmental agencies. But as the evidence from Thailand and to a lesser degree China and Indonesia demonstrate, openness to trade and investment and to new ideas by themselves have not been sufficient to get governments and firms to begin to engage in cleaning up the environment.

Command-and-Control Agencies Bring Change

Real, lasting industrial environmental improvement has depended on governments building and sustaining strong, competent command-and-control environmental agencies with the legal authority to impose significant duties on polluters. For the most part, these agencies have been modeled on the U.S. EPA. Where such agencies have been created (in Malaysia, Singapore, and Taiwan), firms and industrial plants have invested in pollution control, and ambient air and water quality have improved significantly. Where no such agency has been formed (in China, Indonesia, and Thailand), environmental improvements have been more limited (China), sometimes meager (Indonesia), or nonexistent (Thailand), as have improvements in ambient environmental quality.

This finding is important for four reasons. First, some have suggested that traditional command-and-control environmental regulation, as practiced in the OECD countries, will not work in East Asia or that it will take too much time and resources to make it work. My findings strongly suggest otherwise. In fact, they suggest that building such agencies has been the sine qua non of industrial environmental improvement and of improvements in ambient environmental quality in the East Asian NIEs. Second, others have suggested that East Asian governments have been better at promotional policies than they have been at regulatory and policing policies. Because of this, they have doubted that governments in this region would be able to integrate policing policies with promotional policies. Again, my findings strongly suggest that it has not been difficult for governments that depended on promotional policies to turn to policing policies to clean up the environment. In fact, the governments of Malaysia, Singapore, and Taiwan did just this. Third, still others have suggested that because of political, social, cultural, and institutional differences between countries belonging to the OECD (particularly the United States) and countries in East and Southeast Asia, it might be difficult (if not impossible) to simply transfer OECD-style industrial pollution control experiences to Asia. My findings strongly suggest otherwise, particularly for Singapore and Taiwan, and to a lesser degree even for Malaysia. But the ability to do so has depended on the degree of embedded autonomy in governments and on the structure of political institutions rather than on any supposed inability to shift from promotional to policing relationships with business.

The fourth reason, as still others have argued, is that because of the dependence of industrial firms in East Asia on plant and equipment from the OECD countries, industrial environmental improvement in East Asia will crucially depend on the degree to which nonregulatory policies promote the adoption of newer, cleaner investment. In this view, these economies might be able to leapfrog over costly command-and-control policies by maintaining openness to trade and investment; encouraging public disclosure of environmental performance; promoting newer, cleaner investment; and relying more heavily on market-based instruments. This should make it possible for governments in these economies to avoid the costly environmental mistakes made by the OECD countries, take advantage of environmental policy innovations now emanating from the OECD, and meet ambient environmental quality objectives at lower costs. The experience in Thailand suggests otherwise. Among the East Asian NIEs, it was the one that attempted a leapfrogging strategy by relying on pollution prevention, a law mandating a community's right to know, and external pressures from ISO 14000. There is no evidence that any of this has worked. This suggests that the success of these new initiatives may well depend on the existence of a regulatory floor provided by a capable command-and-control regulatory agency. That has been the case in the OECD countries as well. In addition, although policies in these economies are more open than elsewhere, and although some have experimented with public disclosure, virtually none of the governments in this region has actively promoted the adoption of newer, cleaner investment. This is not to say that none of the new investment is cleaner, for surely some of it is. Rather, it is to say that governments in the region, except for Thailand's, have not relied on leapfrogging strategies to reduce pollution. And when they have relied on such strategies, the strategies have not worked.

Embedded Autonomy Increases Effectiveness

Embedded autonomy contributes to the effectiveness of command-and-control environmental regulations. In fact, it appears to be crucial to success and one of the hallmarks of East Asia's experience with controlling industrial pollution. Experiences in China, Malaysia, and Singapore demonstrate how embedded autonomy facilitates industrial environmental improvement. In Singapore, intimate contacts between promoted firms and promotional agencies such as the Economic Development Board (EDB) and the Jurong Town Corporation (JTC) made it easier for Singapore's Ministry of the Environment (ENV) to implement its command-and-control policies. Implementation was facilitated by the fact that promotional privileges from the EDB and JTC were conditioned on promoted firms meeting the ENV's emissions standards. It was also facilitated by the pragmatic bureaucratic culture of the

EDB and the JTC, which filtered down to bureaucrats in the ENV. Officials of the ENV knew that their job was to reduce pollution without significantly slowing economic growth, exports, or the inflow of direct foreign investment. They knew they could not be too tough on polluters. This led them, among other things, to identify international best-practice pollution control technologies, to rely on these technologies in setting emissions standards, and to make information on the technologies available to their business clients.

Embedded autonomy was also important in Malaysia, where the Department of the Environment (DOE) worked closely with a premier industrial policy agency, the Palm Oil Research Institute of Malaysia, to identify pollution technologies that made it possible for the DOE to calibrate emissions standards to the "best available technology not entailing excessive cost." And embedded autonomy was important in China, where the State Environmental Protection Administration (SEPA) and city-level environmental protection bureaus learned to work with traditionally powerful economic agencies to identify cost-effective ways for cities to reduce pollution. In each of these cases, the information and trust gained by having the autonomy of regulatory officials embedded in close relations with those in industrial policy agencies and the private sector facilitated environmental cleanup. But just as important, where embedded autonomy was lacking, as in Thailand and to a lesser degree Indonesia, little industrial environmental improvement occurred.

Pressure Can Make a Difference

Public and community pressure can—when it is followed by the creation of a strong command-and-control environmental agency—make an environmental difference, even in what might be considered authoritarian regimes. This is most clear in Taiwan, where democratization spawned an environmental protest movement that the government recognized it could not ignore. Because of this, the government built and sustained a more and more effective command-and-control agency that imposed duties on polluters.

Similarly, the government of semidemocratic Malaysia responded to growing public complaints about pollution, particularly that from CPO mills, by enacting new environmental legislation and building an effective command-and-control agency. There is also evidence that local receptivity by public officials to complaints about pollution in authoritarian China have made local political leaders such as mayors more prone to actively participate in SEPA's public disclosure program. But as was noted above, environmental progress in China has been much slower than in Malaysia, Singapore, and Taiwan.

Nevertheless, in each of these cases, domestic public pressure appears to have increased the "bargaining" hand of government officials who had longstanding relationships with important actors in the private sector. This, no doubt, made it easier for those officials to get industrial firms to take the need to clean up the environment more seriously than they might have otherwise. But public and community pressure (or the absence of it) does not always matter. In Thailand, where democratization led to the capture of the state by business interests, weak and short-lived coalition governments, and a governmental institutional structure that favors delay and the underprovision of public goods, public pressure has not contributed to more effective environmental management. Moreover, in Singapore environmental improvement was led from above and occurred despite the almost complete absence of public and community pressure to clean up.

Policy Choices for Cleaner Shared Industrial Growth

How might governments in East Asia, and in the developing world more broadly, use these findings to design and implement successful industrial pollution management policies?[2] My answer to this question is based on three key premises. First, most low-income countries in East Asia (and elsewhere in the world) are still in the early stages of their industrial revolutions. Most of the industrial stock that will be in place 25 years from now is not on the ground today.[3] What this means in practice is that when growth accelerates, as it surely will, these low-income economies will likely see the most prodigious expansion of industrial activity in their histories. This is both a threat to sustainability and an opportunity to shape, at an early stage, the water, energy, materials, and pollution intensity of this trajectory of urban–industrial growth. If actions are taken now, there is a once-in-a-lifetime opportunity to achieve a substantially lower, more sustainable growth trajectory.

This opportunity for the second-tier (China, Indonesia, Malaysia, the Philippines, and Thailand) and third-tier (Cambodia, Laos, Papua New Guinea, and Vietnam) East Asian NIEs is significantly different from that faced by the OECD countries or the first-tier NIEs (Hong Kong, Singapore, South Korea, and Taiwan) when they launched their environmental programs in the 1970s and the 1980s. Then, the problem was not how to make the new industrial capital stock cleaner but rather how to retrofit a large existing capital stock with end-of-pipe controls to reduce emissions after they were produced.

The second premise for my answer, as O'Connor (1994) noted, is that the latecomer status of the East Asian NIEs within the global economy is an important context for policy response. This is particularly the case with respect to technology, where the NIEs have access to an array of environmentally advanced technologies developed within the OECD economies.

There is now considerable evidence that in many industries newer plant and equipment developed mainly within the OECD economies tends to be cleaner than existing plant and equipment (Wheeler and Martin 1992; Christensen et al. 1993; Nelson 1994; Arora and Cason 1995; Greiner 1984). This means that it is now, or soon will be, technically and economically possible for manufacturers in the low-income East Asian NIEs to import, operate, adapt, modify, and innovate on an industrial capital stock of process technologies that will tend to be cleaner simply because they are newer.[4] The policy challenge is to promote the selection and use of these cleaner technologies in new industrial investment.

The third premise is that the increased openness of the economies and polities in the low-income East Asian NIEs should enhance the demand for more sustainable outcomes. As the case studies in this book have demonstrated, open economies are already exposing manufacturers in the second-tier NIEs to more and more kinds of environmental market pressure. Sometimes this pressure takes the form of green consumerism, eco-labeling, and the greening of supply chains. Sometimes it takes the form of new international voluntary environmental management standards (ISO 14000) or of industry codes of conduct (e.g., the chemical industry's Responsible Care Program) (Roht-Arriaza 1995). And sometimes it takes the form of corporate disclosure and accountability (e.g., the rapid growth of corporate environmental reporting) (Ditz and Ranganathan 1997; UNEP 1997). Because the miracles of manufacturing growth in East Asia are predicated on exporting manufactures to OECD countries, this external market pressure will only increase and spread over time. It is highly likely that the successful developing-country exporters of manufactures in low-income East Asia will learn to meet these environmental market requirements in the same way that their high-income counterparts learned to meet the industrial-country buyers' price, on-time delivery, quality, and packaging requirements (Keesing 1988).

As the case studies also have demonstrated, the experience of Thailand aside, more and more open polities reinforce this external market demand for sustainability. Citizens, communities, and organized groups in civil society in each of the NIEs, save perhaps for Singapore, are placing increased pressure on the government and private sector to reduce the intensity of industrial production in relation to energy, materials, and pollution. It is now commonplace for environmental agencies to respond to the community's complaints about polluters. It is equally commonplace for governments, communities, and polluters to negotiate pollution abatement agreements. Just as important, polluters are under increasing pressure to compensate communities for environmental damage to crops, fisheries, drinking water supplies, and human health. In some places, such as China and Indonesia, public-sector environmental agencies are taking advantage of

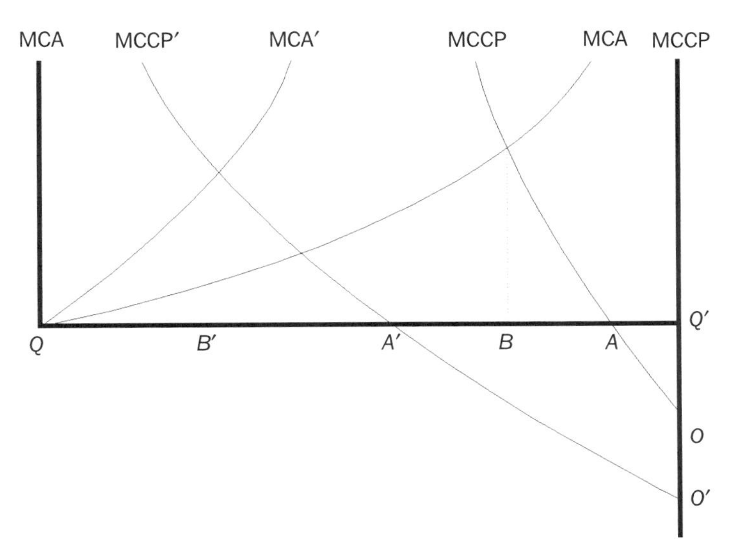

Figure 7-3. Cost-Effective Pollution Reduction: Clean Production versus Pollution Abatement

Note: The left-hand vertical axis represents the marginal dollar cost of pollution abatement. The right-hand vertical axis represents the marginal dollar cost of pollution prevention. The horizontal axis represents either an absolute reduction in pollution or a percentage reduction in pollution to meet ambient environmental standards. The MCA curves, likewise, represent the marginal cost of abatement; the MCCP curves, the marginal cost of cleaner production.

Source: Rock and others 2000; used with the permission of Greenleaf Publishing.

public and community pressure to devise low-cost enforcement strategies that take advantage of the concern of firms for their (environmental) reputations. These pressures will also no doubt spread to the low-income East Asian economies as their growth accelerates.

How might governments in the first-, second-, and third-tier East Asian NIEs (and those elsewhere in the developing world) take advantage of these three premises? The policy choices open to governments can best be demonstrated by reference to a simple diagram (Figure 7-3). Let line *QQ′* in Figure 7-3 equal a desired reduction in the pollution of industrial production for a firm, industry (sector), or economy. Line *QQ′* might measure either an absolute reduction in pollution (in pounds of pollution) or a percentage reduction needed to sustain a given level of ambient environmental quality.[5] The left vertical axis measures the marginal dollar cost of reducing pollution (the marginal cost of abatement, MCA) through traditional postpollution abatement (end-of-pipe expenditures). Curve MCA as drawn (rising from left to right) reflects the traditional rising marginal cost of abatement associated with increasing reductions in pollution through postpollution treatment. The right vertical axis measures the marginal dollar cost of reducing pollution by reducing the energy, water, and materials used in industrial production. This is often referred to as pollution prevention, cleaner production, or what industrial ecologists call "dematerialization" (Ayres and Ayres 1996; Ausubel 1996; Frosch 1996; Graedel and Allenby 1995; Warnick et al. 1996). This curve is labeled MCCP to refer to the marginal cost of cleaner production. It also is reflected in a rising (but from left to right) of the marginal cost of reducing pollution through cleaner production.

There are several important differences between the MCA curve and the MCCP curve. First, to reiterate, MCA reduces pollution by treating pollution after it has occurred, whereas MCCP prevents pollution by reducing the intensity of energy, water, and materials use by substituting less polluting inputs for more polluting inputs; improving energy-, water-, and materials-use efficiencies; and recycling energy, water, and materials. Normally, these cleaner production alternatives are brought about by some combination of better housekeeping practices, minor process modifications, and fundamental technical innovation in industrial production processes. Because of this, reductions in pollution achieved by lowering use intensities are different from those achieved by abating pollution through end-of-pipe treatment.[6] For one, end-of-pipe treatment always increases costs, unlike intensity-reduction activities.[7] This is depicted in Figure 7-3 in the MCCP curve, with an origin that lies below the zero axis. This part of the curve (represented by *OA* and area *OQ'A*) reflects declines in pollution that can be attributed to declining use intensities that "pay." Second, end-of-pipe treatment is almost always a derivative of environmental regulatory policy. Although intensity reductions can flow from regulatory policy, they can also flow from changes in the relative prices of inputs; industrial and investment policies; and the pace, pattern, and rate of diffusion of energy-, water-, and materials-saving technological change. This means that intensity reduction need not be driven solely by regulatory policy. Understanding this and appreciating how regulatory and other policies can reinforce these effects is critical to the design of cost-effective public policies aimed at reducing use intensities. One example of this should suffice.

In the context of the second- and third-tier East Asian NIEs, dematerialization and pollution prevention effects that pay might well represent declines in energy-, water-, and materials-use intensities associated with new (and cleaner) investment. Given the volume of expected new investment relative to the size of the existing industrial capital stock in the second- and third-tier NIEs, these effects could be substantial. This suggests that the governments of these NIEs might consider policies for industry, investment promotion, and technology that encourage firms and plants to adopt and rapidly diffuse cleaner technologies. They might also consider implementing price policies for energy, water, and materials that encourage individual plants, firms, and industries to economize in their use of these inputs.

For heuristic purposes, assume an MCCP given by the curve *OA*MCCP and a marginal cost of abatement given by the curve MCA. Under these conditions, the most cost-effective strategy for reducing pollution in a plant, firm, industry, or economy by *QQ′* requires reductions in pollution through end-of-pipe control by *QB* and reductions in pollution through decreases in energy-, water-, and materials-use intensity by *BQ′*. Note that, as drawn, most of the reduction in pollution comes from conventional end-of-pipe control. Five questions must be asked about this outcome:

- What environmental regulatory policies contribute to this outcome?
- What role do nonregulatory policies have in promoting this outcome?
- Is the outcome depicted the most cost-effective way to reduce pollution and intensities of energy, water, and materials use?
- If the outcome is not the most cost-effective, what might an alternative set of cost-effective policies look like, such as that depicted by outcome *QB′*, *B′Q′*?
- What policy options can be used to encourage individual plants, firms, industries, and economies to strive for continuous "beyond-compliance" environmental improvements (to continually increase *QQ′*)?

Regulatory Policies

If experiences in the OECD and the first-tier East Asian NIEs are any guide, environmental regulatory regimes in the second- and third-tier East Asian NIEs will initially be developed to impose legal limits on emissions from major point sources. Subsequently, those regulatory agencies will learn how to encourage facilities and firms to prevent pollution before it occurs, and then they will learn how to reward firms for continuous beyond-compliance environmental performance (Davies and Mazurek 1996, 1998). In terms of Figure 7-3, policies that impose legal limits on emissions (most often referred to as command-and-control policies) work on MCA; pollution prevention policies work on MCCP; and policies that reward plants, firms, and industries for continuous beyond-compliance performance affect *QQ′*. Initially, the second- and third-tier NIEs' regulatory systems will be asked to protect public health. Most likely, this will be achieved, as it has been in the first-tier NIEs, by establishing clear, consistent ambient environmental standards that are in turn linked to discharge limits for individual facilities.

Once ambient and discharge standards are set, regulators must ensure compliance. Because marginal postpollution abatement costs (MCA in Figure 7-3) are initially low in most industries, most of the second- and third-tier East Asian NIEs are likely to initially focus their compliance efforts on end-of-pipe treatment. A number of the NIEs, particularly Malaysia, Singapore, and Taiwan, have already gone a long way toward doing just this. The other NIEs still need to strengthen the legal authority and institutional and technical capabilities of regulatory agencies. At the same time, the second- and third-tier NIEs can learn from the experiences of the OECD countries and of the first-tier NIEs and thus can take advantage of many emerging regulatory opportunities.

These new opportunities include greater public disclosure of environmental performance, firm- and plant-level flexibility in how performance goals are met, and increased use of market-based instruments, such as pollution taxes and tradable permits. What is important to note is that these new policies will need to be complements to, not substitutes for, basic command-and-control policies that essentially require major point sources to abate pollution by investing in end-of-pipe control equipment. The major effect of this shift toward greater flexibility in how performance goals are met and toward the use of market-based instruments is that these changes make it possible for plants and firms to meet imposed emissions standards at lower abatement costs. That is, they push MCA in Figure 7-3 down and to the right.

Environmental regulatory agencies in the second- and third-tier East Asian NIEs will also likely be asked to reduce industrial pollution through prevention and clean production (the MCCP curve of Figure 7-3, as has already occurred in China and Taiwan). Crucially, this involves correcting the market, policy, and coordination failures that discourage firms from searching for and adopting production practices that lower the intensity of energy, materials, and water use, and of waste.

There are several reasons that policy failures discourage firms from searching for and adopting cleaner production alternatives to pollution intensity reduction. First, traditional command-and-control, technology-based industrial–environmental management systems by themselves tend to favor end-of-pipe pollution reduction strategies over clean production strategies. Because technology-based standards underlying existing command-and-control, industrial–environmental management systems identify the range of pollution reduction possible with the best available end-of-pipe technologies, they are easier and less risky for both regulators and polluters. This biases pollution reduction strategies in an end-of-pipe direction. The use of market-based instruments reinforces this shift. Assuming no change in *O'A'*MCCP', this biases cost-effectiveness toward more reduction by abating pollution after it has occurred.

If, in addition, markets for cleaner production are characterized by information and coordination failures or high risks, and high transaction and learning costs, *O′A′*MCCP′ may be higher and to the right of the existing *O′A′*MCCP′ curve in Figure 7-3. This reinforces the end-of-pipe policy bias. There are several reasons why clean production markets tend to be characterized by information and coordination failures or high risks and transaction and learning costs. To begin with, implementing a firm- or plant-level, clean-production strategy for industrial–environmental management raises several new problems for manufacturing firms and plants. Substitution of a less toxic for a more toxic input may either be perceived to change or actually change the quality of the final product (Laughlin and Corson 1995, 11). Even though it might pay to make this substitution, firms may be unwilling to take the risk of a negative customer reaction to this "new" final product. The same might be said about basic process modifications that pay. In addition, before firms make these switches, they may have to invest scarce managerial and engineering time and even scarcer capital to identify clean production alternatives (Kiesling 1994, 15). Unless these expenditures have known or expected payoffs that are better than the alternatives, firms may be reluctant to make them (Panayotou and Zinnes 1994). That is, it may simply be prudent to stick with well-known end-of-pipe abatement alternatives.

If improved command-and-control policies in the second- and third-tier East Asian NIEs end up biasing industrial–environmental management strategies in an end-of-pipe direction—and if risks, information failures, and learning costs end up undervaluing the benefits of clean production—the governments of these economies should intervene to correct these policy and market failures. This is precisely what regulatory policies that promote cleaner production do. Information, technical assistance, and demonstration projects about pollution prevention opportunities are designed to overcome information failures. Tax breaks (e.g., accelerated depreciation for cleaner production investments) and subsidized loans are meant to level the playing field between pollution abatement alternatives and cleaner production alternatives that reduce the intensity of pollution, energy, and materials use. They are also meant to compensate firms for the risks and learning costs associated with cleaner production alternatives.

If these programs are successful in overcoming policy, market, and information failures (and high transaction and learning costs), the real marginal cost of abatement in Figure 7-3 will be given by MCA′, and the real marginal cost of cleaner production will be given by *O′*MCCP′. With this, three important differences result. First, the range of pollution prevention or energy-, water-, and materials-intensity reduction activities that pay expands from area *OAQ′* to area *O′A′Q′*. This provides more win–win opportunities for polluters. It may also convey competitive advantages to firms that shift in this direction (Porter and van der Linde 1995). Second, cost-effective pollution reduction requires more clean production (an increase in energy, water, and materials intensities reduction from *B* to *B′*) and less end-of-pipe expenditure (a reduction from *B* to *B′*). Third (except if the real *O′A′*MCCP′ is less than the real MCA′ for all levels of pollution reduction), firm- and plant-level cost-effective industrial–environmental management requires identifying the optimal combination of end-of-pipe and clean production.

Clearly, initiatives pursued under the first two environmental regulatory goals described above (protecting public health and supporting clean production) will contribute toward the goal of cleaner industrial growth. But it is likely that the challenge of cleaner growth will place at least one additional new demand on environmental regulatory systems in the second- and third-tier East Asian NIEs. That is, environmental regulatory agencies in these economies will be called on to support a continuous beyond-compliance performance orientation (that continuously increases *QQ′* in Figure 7-3) on the part of firms, industries, and economies. Crucial to this will be improvements in the quality and quantity of information available on the environmental performance of each. It should by now be clear that information is a powerful policy tool.

The case studies in this book reveal the substantial evidence that the development and disclosure of information does, in fact, promote improved environmental performance by industrial firms in East Asia. As Chapter 4 on Indonesia demonstrated, the environmental management agency there used a color-coded environmental rating and disclosure program that has improved the performance of large water polluters. As Chapter 5 showed, China's sustainable-cities index program, which annually rates, ranks, and publicly reports on the performance of its major cities, appears to be influencing the location of industrial activity, the rate of growth of urban infrastructure, and plant-level investments in pollution control.

Nonregulatory Policies

Several researchers (World Bank 1997a, 1994a; Wheeler and Martin 1992) have suggested that newer industrial plants and equipment in industrial countries tend to be cleaner than existing plants and equipment in East Asia. Because manufacturers in East Asia are dependent on industrial-country firms for plants, equipment, and technology, it may be technically and economically possible for them to import, adopt, adapt, modify, and innovate on an industrial capital stock that will be cleaner simply because it is newer. Given the expected increase in the size of the industrial capital stock in the second- and third-tier East Asian NIEs during the next 25 years, this could be an important avenue for cleaner industrial growth. Some (World Bank 1997a) have suggested that because of the openness of the East Asian countries to trade, foreign investment, and foreign technology, this will happen almost automatically.

What are the implications of this possibility for a cost-effective cleaner shared industrial growth outcome, as *QB′*, *B′Q′* depict in Figure 7-3? If openness is sufficient to promote a cleaner industrial capital stock, the effect of openness will be to push *OMCCP* down and to the left so that it moves toward *O′MCCP′*. This results in more pollution reduction through decreasing energy, water, and materials use and less through postpollution abatement. This suggests large win–win effects for the environment and the economy. But it is important to ask if this possibility is inevitable or if it is dependent on other policies. If it is dependent on other policies, it is important to identify those policies.

There are several reasons to suspect that openness, by itself, may not be sufficient to generate win–win outcomes like *QB′*, *B′Q′* in Figure 7-3. First, win–wins will be less likely when "new" investment consists of older and dirtier industrial capital. Second, win–wins will be less likely if policies elsewhere in the economy discourage the efficient use of energy, water, and materials. Third, even if new investment is cleaner and resource pricing policies are efficient, unless firms have the capacity to manage plants and equipment efficiently, they may not be able to achieve the cost-effective reduction outcomes represented by *QB′*, *B′Q′*.

What do we know about each of these? To begin with, there is little doubt that some of the "new" investment in the second- and third-tier East Asian NIEs consists of older and dirtier capital in sunset industries. Several of the first-tier NIEs (Singapore, South Korea, and Taiwan) have encouraged the exporting of low-technology, labor-intensive industries such as textile dyeing, leather making, and electroplating to the second-tier (China, Indonesia, the Philippines, and Thailand) and third-tier (Cambodia, Laos, and Vietnam) East Asian economies. Some (Hsiao 1999, 44; Rock 1996b) have suggested that this is the natural outcome of shifting comparative advantage. This suggests that openness alone might just as easily promote dirtier industrial outcomes. This tendency can be and has been exacerbated by inappropriate pricing policies for energy, water, and other materials in some of the NIEs. Sometimes, as in China, energy price policy favors dirty over clean fuels. Sometimes, as in Indonesia, energy prices are kept well below international prices. Similarly, water and other materials (e.g., wood and primary metals) are also often underpriced.

How do importing older and dirtier capital equipment and underpricing energy, water, and materials affect the pollution reduction outcomes depicted in Figure 7-3? Importing older and dirtier equipment has at least two effects. First, it forces firms and plants to rely on end-of-pipe treatment. Second, it may provide opportunities for plants to engage in better housekeeping practices and minor process changes that reduce the intensity of energy, water, and materials use. But the extent to which each plant engages in these activities will depend on the extent to which regulatory policies encourage both end-of-pipe treatment and cleaner production. Its engagement will also depend on price policies for energy, water, and materials. If regulatory policies emphasize end-of-pipe treatment and price policies discourage the efficient use of each, outcomes will look more like *QB*, *BQ'* rather than *QB'*, *B'Q'*. But if regulatory policies encourage clean production alternatives, as well as end-of-pipe treatment and energy, and prices reflect international prices, then pollution reduction could move more toward outcomes like *QB'*, *B'Q'*.

That being said, before industrial firms in the second- and third-tier East Asian NIEs can take advantage of opportunities for either end-of-pipe or cleaner production, they must possess the technical know-how and have the capability to efficiently manage their plants, equipment, and technology. If industrial firms lack the capability to do these things, there may be significant limits to their ability to respond to regulatory, economywide, and industrial-policy incentives designed to push them in a direction that lowers the intensity of pollution, energy, water, and materials use. A lack of capability in these areas might also limit the ability of firms to take advantage of new, cleaner imported technologies.

What do we know about the capability of firms in the East Asian NIEs to manage production efficiently, to improve production efficiency, and to carry out technical change? We know three kinds of principles. First, there is enormous variability in the technological capabilities of firms (Roberts and Tybout 1996; Kim 1997; Felker 1998; Rock 1995, 1999). Capabilities vary by economy, firm size, sector, and ownership. Firms in Northeast Asia appear to be more capable than their counterparts in Southeast Asia (Kim 1997; Felker 1998). Large firms appear to be more capable than small firms (Lall 1992, 169). Firms in supplier-dominated capital-goods sectors (e.g., textiles) appear to be more capable than in either scale-intensive sectors (e.g., automobiles and aircraft) or science-based sectors (e.g., chemicals or electronics, which need a large capacity for reverse engineering) (Bell and Pavitt 1992, 265). Firms engaged in joint ventures with large foreign firms appear to be more capable than domestically owned ones (Harrison 1996, 167–173).

Second, because the acquisition of these capabilities is mostly tacit (i.e., it can only be gained from direct experience), their variability also depends on a firm's willingness to invest in learning by doing (Bell and Pavitt 1992, 262)—which also appears to be enormously variable. Moreover, this willingness is strongly influenced by country policies. A stable, high-growth environment appears particularly conducive to firms' willingness to invest in technological capability (Lall 1992, 169). Export-oriented industrialization policies that require firms to reduce costs, raise quality, and introduce new products help. When trade policy is tethered to lucrative export incentives, it can be a powerful stimulus to technical capability building within firms (Rhee et al. 1984; Kim 1997). State policies that favor and reward local-firm technical capacity acquisition over reliance on foreign capital (foreign direct investment) can and have reinforced these effects (Mardon 1990).

Third, because there are significant externalities in the accumulation of production and technological capabilities, government policies are needed to speed the process by which firms acquire new technical capabilities and diffuse them throughout the economy. The experiences of the first-tier East Asian NIEs suggest that two distinct sets of issues affect the speed with which firms acquire these capabilities. The first concerns the influence of government policy on firm size. The second concerns the need for governments to invest in the provision of public goods that speed the acquisition of technical capabilities in industrial firms.

With respect to the size of firms, two distinct patterns have emerged. In South Korea, one early aim of government policy was to promote the development of very large firms (*chaebols*) that could internalize, and hence appropriate, many of the externalities associated with technological learning (Lall 1992, 176; Jones and Sakong 1980). When this was combined with stable and high growth, export orientation, and an administrative structure that rewarded performance, the consequences for the acquisition of technical capabilities were enormous (Kim 1997). Government support for the development of equally large industrial conglomerates in Indonesia, Thailand, and Malaysia suggests that something similar may be at work in these countries (Rock 1995, 1999, 2000b). There is one other benefit of government policies promoting the development of large diversified industrial conglomerates: Some of those firms are likely to become leading firms. As the experiences of the OECD countries show, leading firms appear to be particularly susceptible to incentives designed to reward continuous beyond-compliance performance (i.e., to get them to reduce pollution by more than *QQ′* in Figure 7-3).

Alternatively, industrial development policy in Taiwan promoted the development of small rather than large firms. Because none of these in any industry was capable of internalizing the externalities associated with all facets of the acquisition of technical capabilities, much of this was done either in government-funded industrial technology research institutes or in joint programs of the public and private sectors coordinated by the government (Lall 1992, 176; Wade 1990). When this happens, it is not surprising that the public rather than private sector takes the lead in clean production and superior environmental performance.

Beyond this, public-sector investments in national technical capability building also matter (Nelson 1993). As the experiences of South Korea and Taiwan demonstrate, large investments in literacy, secondary education, and tertiary education, particularly engineering training, make it easier for firms to acquire technical capabilities (Tan and Batra 1995, 1, 7). By providing information (including information on cleaner technologies), testing materials, inspecting and certifying quality-control standards (including ISO 14000), and calibrating instruments of measurement, a technology infrastructure facilitates the acquisition of technical capabilities, particularly in small and medium-sized enterprises (SMEs).

In the first-, second-, and third-tier East Asian NIEs, then, what are the policy implications of all of this for diffusion among firms of technological capability in pollution reduction and in reducing the intensity of energy, water, and materials use? There are two answers to this question. First, policies that promote firm-level technical learning and capability acquisition are likely to be good for intensity reduction. These policies should make it easier for firms to engage in better housekeeping practices and minor process innovations that prevent pollution. And these policies should make it possible for firms to "stretch" existing plant and equipment by substantially modifying it to reduce pollution, energy, water, and materials use. They should also make it easier for firms to evaluate the intensity of "new" imported plant, equipment, and technology. Each of these lowers the marginal costs of cleaner production (in Figure 7-3, MCCP shifts to the left) and contributes to more pollution reduction by increasing use efficiencies.

Second, because the reduction of pollution, energy, water, and materials intensity is or will be a relatively new activity for industrial firms (particularly in the second- and third-tier East Asian NIEs), industrial firms are likely to need industry- and technology-specific information and specialized technical training on how to do this. This is just the kind of information and specialized training that institutions that are part of the national technology infrastructure (e.g., industrial technology institutes or standards agencies) are good at providing. They should be encouraged to provide such information and training to overcome information failures and the high transaction costs associated with reducing intensities of use. This is most likely to be true for SMEs where existing SME and multinational-corporation linkage programs aimed at technologically upgrading SMEs might well be modified to include making the corporations' supply chains more environmentally responsible (Battat et al. 1996).

Conclusions: Tailoring Policies to Economies

The arguments in this chapter suggest that getting policies right in three discrete but overlapping policy arenas—environmental regulation; trade and resources pricing; and industrial development, investment promotion, and technology—are critical to the success of industrial pollution reduction strategies in the East Asian NIEs. How might individual economies and subregions in East Asia, such as the Association of Southeast Asian Nations

(ASEAN), use these insights to design and implement cost-effective intensity-reduction policies?

To begin with, virtually all of these economies can gain by pricing energy, water, and materials closer to their real scarcity values. As a start, they can gain by removing distortions and allowing prices for these inputs to move closer to traded or international prices or to the long-run marginal costs of production. Each of these economies can also gain by maintaining and increasing openness to trade, foreign direct investment, and foreign technology and by policies that encourage firms to engage in high-speed technological learning and capacity building. Public investments in national technological capacity building and incentives that reward individual firms for engaging in high-speed technological learning should also help firms move toward cost-effective pollution reduction. Beyond this, policies need to be tailored to take advantage of differences in existing conditions in each of the East Asian NIEs.

In tailoring policies to each East Asian NIE, three groups of economies with varying patterns of conditions and needs need to be delineated. The first group—the first-tier economies, Malaysia, Singapore, South Korea, and Taiwan—has relatively strong command-and-control environmental agencies, economies that are nearing the end of their industrial revolutions, and firms with strong technical capabilities. The second group—the second-tier economies, China, Indonesia, the Philippines, and Thailand—has much weaker environmental protection agencies, economies that are in the midst of their industrial revolutions, and firms with weaker technical capabilities. The third group—the third-tier economies, Cambodia, Laos, Papua New Guinea, and Vietnam—has extremely weak environmental protection agencies, economies that are at the beginning of their industrial revolutions, and firms with extremely limited technical capabilities.

The first-tier NIEs face four problems and opportunities. To begin with, because these economies are nearing the end of their industrial revolutions, their intensity of pollution, energy, water, and materials use is likely to grow less fast than their income. Moreover, most of their industrial capital stock that will be in place 25 years from now is already functioning. Because of this, and because these economies have relatively successful command-and-control environmental agencies, their cleanup is either just about complete (in Singapore) or well on the way to being complete (Malaysia and Taiwan).

Because the environmental agencies in these first-tier NIEs are command-and-control oriented, their pollution reduction has been biased in an end-of-pipe direction. As we know from the experience of the OECD countries and Singapore, there are rapidly diminishing returns to this strategy (Ling 1994). As ambient environmental standards and facilities-specific emissions standards are tightened, firms in these economies will be forced to move further up the marginal cost-of-abatement curve. This will undoubtedly create pressures, as it did in the OECD countries, on regulators to ease up on the regulated community. Because of the close relationship between business and government in these economies, this could contribute to regulatory reversals. To counter this, regulatory agencies in these economies need to develop market-based instruments, pollution prevention, and continuous beyond-compliance performance complements to command-and-control policies. This means that regulatory agencies in these economies are likely to be particularly open to policy initiatives that work on MCCP in Figure 7-3 (prevent pollution) and expand *QQ′* beyond what regulations require (reward superior performance). Regulatory agencies in these economies also need to develop stronger relationships with and more support for their actions with political leaders, the public, and the regulated community. This may be necessary to prevent regulatory backsliding. Because publics, communities, and environmental NGOs in these economies tend to be distrustful of governments, this may not be easy to do (Lee and So 1999).

These first-tier economies have made a habit of engaging in high-speed technological borrowing and learning. Therefore, it should be relatively easy for them to do the same in environmental management. Tough, competent regulatory agencies have already contributed to this, at least with respect to end-of-pipe solutions to pollution. Now is the time to extend firm-level learning to cleaner production and superior performance solutions. How this might best be done is likely to vary by economy. In South Korea, where large, vertically integrated conglomerates dominate, much of the new learning is likely to take place within the firm. Thus, policies designed to promote technical environmental learning in cleaner production and superior performance must take account of this. One way to do this is to link corporate leaders and environmental management units in these large firms with their counterparts in leading U.S. firms. In Taiwan, where small firms dominate, the public sector is likely to be the primary conduit for learning about cleaner production and superior performance. This requires working with industrial policy agencies (e.g., the Industrial Development Bureau of the Ministry of Economic Affairs), science and technology institutes (e.g., the Industrial Technology Research Institute), and standards agencies.

Finally, the governments of these first-tier economies are engaged in selective industrial policies that promote the development of indigenous environmental goods and services industries. In each instance, nascent domestic industries producing environmental goods and services are expected to become export oriented. In some economies (Taiwan), government agencies expect this new industry to capture a significant share of the market for environmental goods and services in countries such as Indonesia, the Philippines, and Thailand. It would be unfortunate if firms in this industry and in these economies end up successfully promoting and exporting only end-of-pipe solutions to pollution. To avoid a bias toward end-of-pipe solutions, efforts should be made to ensure that technological capacity building in this industry includes learning about cleaner production and superior beyond-compliance performance policies

The second-tier NIEs face more difficult tasks. For one, their environmental regulatory agencies are much weaker. In some of them (the Philippines and Thailand), these agencies operate without the strong environmental legislation that would empower them to set ambient and emissions standards, monitor performance, and enforce compliance. In others (Indonesia and Thailand), regulatory agencies have no authority to monitor, inspect, or enforce facilities-specific emissions standards. In virtually all of them, regulatory agencies lack both the technical capacity and the resources to manage a national environmental protection program. Weaknesses in protection programs will be exacerbated by substantial and massive increases in industrial output during the next 25 years. This combination of weak environmental protection agencies and large expected increases in output is particularly threatening.

What can and should governments do under these circumstances? First and foremost, a substantial effort must be made to enhance the capabilities of environmental protection agencies to set, monitor, and enforce facilities-specific emissions standards. The experiences of Malaysia, South Korea, Singapore, and Taiwan (as well as of OECD countries) suggest that this effort will take time and resources. In these four NIEs, and in the OECD more broadly, the effort required building the capacity of agencies to implement and manage traditional command-and-control policies. Once this was done, regulatory agencies introduced pollution prevention and superior performance policies. This raises an interesting question. Should the nascent agencies in these NIEs follow this path, or should they try to simultaneously develop command-and-control, pollution prevention, and superior performance policies? Or should they attempt even more innovative alternatives, such as integrated pollution control (Hersh 1996)? Because pollution prevention policies and superior performance policies are complements to and not substitutes for sound command-and-control policies, these economies would probably be best served by initially developing the capacity to manage rigorous command-and-control programs.

What might these agencies do while command-and-control capacities are being built? There is a simple, straightforward answer: As the Chinese and Indonesian case studies indicated, environmental protection agencies need to be both opportunistic and strategic. That is, they need to look for opportunities where they can intervene to make a difference and where they can learn by doing. This suggests taking a problem-specific approach to capacity building. This can mean taking action that either builds on or galvanizes public opinion or community pressure. As Chapter 6 demonstrated, the Department of the Environment in Malaysia took advantage of growing community and public dissatisfaction over unabated pollution from crude palm oil mills to fashion a highly effective intervention strategy that successfully delinked production and exports from water pollution. This included development of a highly productive relationship with a quasi-public, quasi-private research institute for science and technology. As Chapter 4 showed, a local environmental agency in Indonesia did much the same when it used a highly publicized pollution incident to mount a small-scale monitoring and inspection program that worked (also see Aden and Rock 1999). Indonesia's national environmental impact agency, BAPEDAL, went one step further by developing a simple environmental business rating program, PROPER, that relies on public disclosure and shame to get plants to clean up pollution. And as Chapter 5 demonstrated, China's State Environmental Protection Administration developed a similar public disclosure program for cities.

The export orientation of firms in these economies opens an additional opportunity for strategic intervention. There is growing evidence that external environmental market pressure can influence the environmental behavior of manufacturing plants that export. Sometimes this takes the form of improving the environmental responsiveness of supply chains, sometimes it takes the form of international voluntary environmental standards (e.g., ISO 14000), and sometimes it takes the form of industry codes of conduct (e.g., the chemical industry's Responsible Care program). Nascent environmental regulatory agencies in this group of economies can take advantage of the opportunity created by the export orientation of industry by working with industrial policy agencies (ministries of industry, science and technology institutes, and standards agencies) that provide assistance to local firms so they can meet these requirements. This might take the form of cooperation between an environmental protection agency and a national standards agency on developing policies for the ISO 14000 certification of local firms. It might take the form of adding an environmental supply chain to linkage programs between local SME suppliers and multinational buyers. Or it might take the form of a product eco-labeling program developed by environmental protection agencies in cooperation with respected domestic environmental NGOs.

These kinds of partnership programs between environmental protection agencies and industrial policy agencies have three potential advantages. Because the partnerships place some of the implementation burden on others, they limit demands on nascent environmental protection agencies. They also encourage productive relationships between the two kinds of agencies. This can work to the benefit of the industrial policy agencies, particularly as the environmental agencies learn that the industrial agencies can help clients meet some external environmental requirements. Finally, such partnerships actively engage industrial policy agencies in environmental protection.

The third-tier NIEs face the most formidable challenges. These economies are largely agrarian, have very small industrial bases, and have even smaller export-oriented industrial bases. Their current comparative advantage in industry is in low-skill, low-wage, labor-intensive dirty industries such as textile dyeing, leather making, and electroplating. These are relatively footloose industries and the very industries from which other NIEs—particularly Hong Kong, Singapore, South Korea, and Taiwan—are losing comparative advantage. Because of this loss of comparative advantage, many of these industries are relocating to these third-tier economies. Some industries are also moving to other low-wage countries (e.g., Bangladesh, India, and Sri Lanka).

Comparative advantage in dirty industries, high levels of poverty, low levels of education, and great weaknesses in institutional capacity in government generally and in environmental protection in particular provide few obvious opportunities for effective government intervention. These economies might have much to gain from a regional (ASEAN-based) investment code of environmental conduct that binds foreign investors to an agreed-upon set of environmental practices. This could be particularly helpful if investors from elsewhere in East Asia and from OECD countries abided by a set of environmental requirements similar to those of investors' home countries or economies. Export-oriented industrial plants in these economies might also gain from environmentally responsive supply chains and other aspects of the external environmental market, such as ISO 14000 certification and eco-labeling, particularly if they are managed either by foreign investors or donors.

Notes

[1]The term "midwifery" comes from Evans (1995, 13–16), and it refers to government policies aimed at assisting new entrepreneurs or inducing existing entrepreneurs to take on new tasks.

[2]Most of what follows in the remainder of this chapter has been adapted from Rock and others 2000, with the permission of Greenleaf Publishing.

[3]In a pre-crisis study of Indonesia, the World Bank (1994a) projected that 85% of the capital stock that would be in place by 2020 was not yet in place. Even with the current recession and slower, delayed growth, the significance of new investment remains. At an annual growth rate of manufacturing output of 7.25% (just half the growth rate maintained during the 1990s before the crisis), manufacturing output doubles every 10 years.

[4]There are, however, no good data on the extent to which manufacturers are availing themselves of cleaner technology within Southeast Asia. Because most existing policy presumes a 20–30% improvement in energy and materials efficiency simply through the use of newer, cleaner technology, this becomes a critical policy issue.

[5]If the scale of industrial activity increases, the size of QQ' may have to be expanded to sustain a given level of ambient environmental quality.

[6]This is particularly important for some pollutants, such as carbon dioxide, that simply cannot be abated by end-of-pipe technologies.

[7]But not all cleaner production opportunities pay, either.

Bibliography

Actherberg, W. 1996. Sustainability, Community, and Democracy. In *Democracy and Green Political Thought*, edited by B. Doherty and M. de Geus. New York: Routledge, 170–187.

Aden, J., A. Kyu-Hong, and M.T. Rock. 1999. What Is Driving the Pollution Abatement Expenditure Behavior of Manufacturing Plants in Korea? *World Development* 27: 1203–1214.

Aden, J., and M.T. Rock. 1999. Initiating Environmental Behavior in Manufacturing Plants in Indonesia. *Journal of Environment and Development* 8(4): 357–375.

Afsah, S., B. Laplante, and N. Makarim. 1995. *Program-based Pollution Control Management: The Indonesian PROKASIH Program.* Policy Research Department. Washington, DC: World Bank.

Afsah, S., and J. Vincent. 2000. Putting Pressure on Polluters: Indonesia's PROPER program. In *Asia's Clean Revolution*, edited by D. Angel and M.T. Rock. Sheffield, U.K.: Greenleaf Publishing, 157–172.

Amsden, A. 1979. Taiwan's Economic History: A Case of Etatisme and a Challenge to Dependency. *Modern China* 5(3): 341–380.

Anek, L. 1988. Business and Politics in Thailand: New Patterns of Influence. *Asian Survey* 28: 451–470.

———. 1992. *Business Associations and the New Political Economy of Thailand: From Bureaucratic Polity to Liberal Corporatism.* Boulder, CO: Westview Press.

Angel, D., and M.T. Rock, eds. 2000. *Asia's Clean Revolution*. Sheffield, U.K.: Greenleaf Publishing.

Arora, S., and T. Cason. 1995. An Experiment in Voluntary Environmental Regulation: Participation in EPA's 33/50 Program. *Journal of Environmental Economics and Management* 28: 271–286.

Ausubel, J.H. 1996. The Liberation of the Environment. *Daedalus* 125(3): 1–18.

Ayres, R., and L. Ayres. 1996. *Industrial Ecology.* Cheltenham, U.K. Edward Elgar.

Bangkok Post. 1982. Working Together at Last. December 31, 31–32.

Barr, C. 1998. Discipline and Accumulate: State Practice and Elite Consolidation in Indonesia's Timber Sector, 1967–98. Master's thesis, Cornell University, Ithaca, NY.

Barraclough, S. 1984. Political Participation and Its Regulation in Malaysia: Opposition to the Societies (Amendment) Act of 1981. *Pacific Affairs* 53(3): 450–461.

Basri, C., and H. Hill. 1996. The Political Economy of Manufacturing Protection in LDCs: An Indonesian Case Study. *Oxford Development Studies* 24(3): 241–259.

Battat, J., I. Frank, and X. Shen. 1996. *Suppliers to MNCs: Linkage Programs to Strengthen Local Companies in Developing Countries.* Occasional Paper 6, Foreign Investment Advisory Service. Washington, DC: World Bank.

Bell, M., and K. Pavitt. 1992. Accumulating Technological Capability in Developing Countries. In *Proceedings of the World Bank Annual Conference on Development Economics.* Washington, DC: World Bank, 257–281.

Bello, W., and S. Rosenfeld. 1990. *Dragons in Distress: Asia's Miracle Economies in Crisis.* San Francisco: Institute for Food Policy and Development.

———. 1992. *Dragons in Distress.* New York: Food First.

Booth, A. 1989. Indonesia's Agricultural Development in Comparative Perspective. *World Development* 17(8): 1235–1254.

Bowie, A., and D. Unger. 1997. *The Politics of Open Economies: Indonesia, Malaysia, the Philippines, and Thailand.* Cambridge, U.K.: Cambridge University Press.

Brandon, C., and R. Ramankutty. 1993. *Toward an Environmental Strategy for Asia.* Washington, DC: World Bank.

Breslin, S.G. 1996. *China in the 1980s: Centre-Province Relations in a Reforming Socialist State.* New York: Saint Martin's Press.

Bresnan, J. 1993. *Managing Indonesia.* New York: Columbia University Press.

Bruton, H.J., G. Abeysekera, N. Sanderatone, and Z.A. Yusof. 1992. *The Political Economy of Poverty, Equity, and Growth: Sri Lanka and Malaysia.* New York: Oxford University Press.

Bureau of Water Quality Protection, TEPA (Taiwan Environmental Protection Administration). 2001. *Pollution Status of Major and Secondary Rivers in Taiwan: 1981–1999.* Taipei: TEPA.

Callahan, W.A., and D. McCargo. 1996. Vote-Buying in Thailand's Northeast. *Asian Survey* 36(4): 376–391.

Case, W. 1993. Semi-Democracy in Malaysia: Withstanding the Pressures for Regime Change. *Pacific Affairs* 66(2): 183–205.

———. 1995. Malaysia: Aspects and Audiences of Legitimacy. In *Political Legitimacy in Southeast Asia,* edited by M. Alagappa. Stanford: Stanford University Press, 69–107.

Chai-anan, S. 1971. *The Politics and Administration of the Thai Budgetary Process.* Doctoral dissertation, University of Wisconsin, Madison.

———. 1990. Thailand: A Stable Semi-Democracy. In *Politics in Developing Countries,* edited by L. Diamond, J. Linz, and S.M. Lipset. Boulder, CO: Lynne Reiner.

Chan, D.C. 1993. The Environmental Dilemma in Taiwan. *Journal of Northeast Asian Studies* 12(1): 35–56.

Chern, K.S. 1976–1977. Politics of America's China Policy, 1945: Roots of the Cold War. *Political Science Quarterly* 91(4): 631–647.

Chew, C. 1974. River Pollution Puts Paid to Days of Plenty. *New Straits Times*, February 3, 19.

Christensen, K., B. Nielson, P. Doelman, and R. Schellman. 1995. Cleaner Technologies in Europe. *Journal of Cleaner Production* 3(1–2): 67–70.

Christensen, S. 1991. Thailand after the Coup. *Journal of Democracy* 2(3): 94–106.

Christensen, S., D. Dollar, A. Siamwalla, and P. Vichyanond. 1993. *The Lessons of East Asia: Thailand, the Institutional and Political Underpinnings of Growth.* Washington, DC: World Bank.

Chun-Chieh, C. 1994. Growth with Pollution: Unsustainable Development in Taiwan and Its Consequences. *Studies in Comparative International Development* 29(2): 23–47.

Congleton, R.D. 1992. Political Institutions and Pollution Control. *Review of Economics and Statistics* 74(3): 412–421.

Cox, G.W., and M.D. McCubbins. 2001. The Institutional Determinants of Economic Policy. In *Presidents, Parliaments and Policy,* edited by S. Haggard and M.D. McCubbins. Cambridge, U.K.: Cambridge University Press, 21–64.

Cribb, R. 1990. The Politics of Pollution Control in Indonesia. *Asian Survey* 30(12): 1123–1135.

Crouch, H. 1996. *Government and Society in Malaysia.* Ithaca, NY: Cornell University Press.

Dalpino, C. 1991. Thailand's Search for Accountability. *Journal of Democracy* 2(4): 61–69.

Davies, J.C., and J. Mazurek. 1996. *Industry Incentives for Environmental Improvement: Evaluation of U.S. Federal Initiatives.* Washington, DC: Global Environmental Management Initiative.

———. 1998. *Regulating Pollution. Does the U.S. System Work?* Washington, DC: Resources for the Future.

Davis, D.S., R. Kraus, B. Naughton, and E.J. Perry. 1995. *Urban Spaces in Contemporary China.* New York: Cambridge University Press.

Ditz, D., and J. Ranganathan. 1997. *Measuring Up: Toward a Common Framework for Tracking Corporate Environmental Performance.* Washington, DC: World Resources Institute.

DOE (Department of the Environment, Malaysia). 1992. *Malaysia: Environmental Quality Report, 1992.* Ministry of Science, Technology, and the Environment. Kuala Lumpur: DOE.

———. 1994. *Malaysia: Environmental Quality Report, 1994.* Ministry of Science, Technology, and the Environment. Kuala Lumpur: DOE.

———. 1997. *Malaysia: Environmental Quality Report, 1997.* Ministry of Science, Technology, and the Environment. Kuala Lumpur: DOE.

———. 1998. *Malaysia: Environmental Quality Report, 1998.* Ministry of Science, Technology, and the Environment. Kuala Lumpur: DOE.

Eder, N. 1996. *Poisoned Prosperity.* Armonk, NY: M.E. Sharpe.

Ehrenfield, J., and N. Gertler. 1997. Industrial Ecology in Practice: The Evolution of Interdependence at Kaolundborg. *Journal of Industrial Ecology* 1(1): 67–97.

Evans, P. 1995. *Embedded Autonomy: States and Industrial Transformation.* Princeton, NJ: Princeton University Press.

Felker, G.B. 1998. Upwardly Global: The State, Business, and the MNCs in Malaysia's and Thailand's Industrial Transformation. Doctoral dissertation, Woodrow Wilson School, Princeton, NJ.

Ferris, R.A. 1993. Aspiration and Reality in Taiwan, South Korea, Hong Kong and Singapore: An Introduction to the Environmental Regulatory Systems of Asia's Four New Dragons. *Duke Journal of Comparative and International Law* 4(1): 125–187.

Findlay, R., and S. Wellisz. 1993. Introduction. In *Five Small Open Economies,* edited by R. Findlay and S. Wellisz. Oxford, U.K.: Oxford University Press, 1–15.

Fiorino, D. 1989. Environmental Risk and Democratic Process. *Columbia Journal of Environmental Law* 6(3): 501–544.

Foo, K.M., L.H. Lye, and K.L. Koh. 1995. Environmental Protection: The Legal Framework. In *Environment and the City,* edited by O.G. Ling. Singapore: Institute of Policy Studies and Times Academic Press, 47–99.

Frosch, R. 1996. Toward the End of Waste: Reflections on a New Ecology of Industry. *Daedalus* 125(3): 199–211.

Gelb, A., and associates. 1988. *Oil Windfalls: Blessing or Curse.* New York: Oxford University Press.

Gillis, M. 1984. Episodes of Indonesian Economic Growth. In *World Economic Growth,* edited by A.C. Harberger. San Francisco: International Center on Economic Growth, 231–264.

Girling, J. 1981. *Thailand: Society and Politics.* Ithaca, NY: Cornell University Press.

Glassburner, B. 1978a. Indonesia's New Economic Policy and Its Sociopolitical Implications. In *Political Power and Communication in Indonesia,* edited by K.D. Jackson and L. Pye. Berkeley, CA: University of California Press, 137–143.

———. 1978b. Political Economy and the Soeharto Regime. *Bulletin of Indonesian Economic Studies* 14(3): 24–51.

Graedel, T.E., and B.R. Allenby. 1995. *Industrial Ecology.* Upper Saddle River, NJ: Prentice Hall.

Greiner, T. 1984. The Environmental Manager's Perspective on Toxics Use Reduction Planning. Master's thesis, Massachusetts Institute of Technology, Cambridge, MA.

Haggard, S. 1990. *Pathways from the Periphery.* Ithaca, NY: Cornell University Press.

Haggard, S., and M.D. McCubbins, eds. 2001. *Presidents, Parliaments and Policy.* Cambridge, U.K.: Cambridge University Press.

Harrison, A. 1996. Determinants and Effects of Direct Foreign Investment in Côte d'Ivoire, Morocco, and Venezuela. In *Industrial Evolution in Developing Countries,* edited by M. Roberts and J. Tybout. New York: Oxford University Press, 163–186.

Hersh, R. 1996. A Review of Integrated Pollution Control in Selected Countries. Discussion paper 97–117. Washington, DC: Resources for the Future.

Hettige, H., M. Huq, S. Pargal, and D. Wheeler. 1996. Determinants of Pollution Abatement in Developing Countries: Evidence from South and South East Asia. *World Development* 24(12): 1891–1904.

Hettige, H., M. Mani, and D. Wheeler. 1998. *Industrial Pollution in Economic Development: Kuznets Revisited.* Working Paper 1876, World Development Research Group. Washington, DC: World Bank.

Hettige, H., P. Martin, M. Singh, and D. Wheeler. 1995. *The Industrial Pollution Projection System.* Washington, DC: World Bank.

Hicken, A. 1998. From Patronage to Policy: Political Institutions and Policy Making in Thailand. Paper presented at Midwest Political Science Association Meeting, April 23–25, Chicago.

———. 1999. Parties, Politics, and Patronage: Governance and Growth in Thailand. Working paper. San Diego: University of California.

Hill, H. 1996a. *The Indonesian Economy since 1966.* Cambridge, U.K.: Cambridge University Press.

———. 1996b. October. Indonesia's Industrial Policy and Performance: Orthodoxy Vindicated. *Economic Development and Cultural Change* 45: 146–174.

Hirsch, P., and L. Lohmann. 1989. The Contemporary Politics of Environment in Thailand. *Asian Survey* 29: 439–451.

Hsiao, M. 1999. Environmental Movements in Taiwan. In *Asia's Environmental Movements: Comparative Perspectives,* edited by Y.F. Lee and A.Y. So. Armonk, NY: M.E. Sharpe, 31–54.

Huff, W.G. 1994. *The Economic Growth of Singapore.* Cambridge, U.K.: Cambridge University Press.

———. 1995. The Developmental State, Government, and Singapore's Economic Development since 1960. *World Development* 23(8): 1421–1438.

Hughes, H. 1993. An External View. In *Challenge and Response: Thirty Years of the Economic Development Board.* Singapore: Times Academic Press, 1–26.

Hui, J. 1995. Environmental Policy and Green Planning. In *Environment and the City,* edited by O.G. Ling. Singapore: Times Academic Press, 13–46.

IDB (Industrial Development Bureau, Taiwan). 1995. *Development of Industries in Taiwan Republic of China.* Taipei: Ministry of Economic Affairs.

———. No date. *Promotion and Accomplishment of Industrial Pollution Prevention and Control in the Republic of China.* Taipei: Ministry of Economic Affairs.

Ingram, J. 1971. *Economic Change in Thailand: 1850–1970.* Stanford, CA: Stanford University Press.

Jahiel, A.R. 1997. The Contradictory Impact of Reform on Environmental Protection in China. *The China Quarterly* 149(March): 81–103.

———. 1998. The Organization of Environmental Protection in China. *China Quarterly* 157(December): 757–787.

Johnson, C. 1987. Political Institutions and Economic Performance: The Government–Business Relationship in Japan, South Korea, and Taiwan. In *The Political Economy of New Asian Industrialism,* edited by F.C. Deyo. Ithaca, NY: Cornell University Press, 136–164.

Johnston, B.F., and P. Kilby. 1975. *Agriculture and Structural Transformation.* New York: Oxford University Press.

Jomo K.S. 1986. *A Question of Class: Capital, the State and Uneven Development in Malaya.* Singapore: Oxford University Press.

Jones, C. 1995. Rice Price Stabilization in Indonesia: An Economic Assessment of the Changes in Risk Bearing. *Bulletin of Indonesian Economic Studies* 31(1): 109–128.

Jones, L., and I. Sakong. 1980. *Government, Business, and Entrepreneurship: The Korean Case.* Cambridge, MA: Harvard University Press.

Kanittha, I. 1996. Levels of Pollution Remain Stagnant. *Bangkok Post*, December 31, 1–3.

———. 1997. Officials Claim Victory over Stench from Industrial Estate. *Bangkok Post*, July 24, 1–3.

Karl, T.L. 1990. Dilemmas of Democratization in Latin America, October. *Comparative Politics* 23(1): 1–21.

Keesing, D. 1988. The Four Exceptions: UNDP—Trade Expansion Program. Occasional paper 2. New York: United Nations Development Program.

Khoo, C.H. 1991. Environmental Management in Singapore. *Environmental Monitoring and Assessment* 19: 127–130.

Kiesling, F. 1994. *Minnesota P2 Planning Survey: Results and Technical Report.* Minneapolis: Minnesota Survey Research Center, University of Minnesota.

Kim, L. 1997. *From Imitation to Innovation.* Boston, MA: Harvard Business School Press.

King, D.E. 1996. Thailand in 1995: Open Society, Dynamic Economy, Troubled Politics. *Asian Survey* 36(2): 135–141.

Kirtada, M., and G.S. Heike. 1995. The Role of NGOs and Near NGOs. In *Environment and the City,* edited by O.G. Ling. Singapore: Institute of Policy Studies and Times Academic Press, 282–303.

Klein, J.R. 1998. The Constitution of the Kingdom of Thailand: A Blueprint for Participatory Democracy. Working paper 8. San Francisco: Asia Foundation.

Koe, L.C.C., and M.A. Aziz. 1995. Environmental Protection Programs. In *Environment and the City,* edited by O.G. Ling. Singapore: Institute of Policy Studies and Times Academic Press, 200–220.

Komin, S. 1993. A Social Analysis of the Environmental Problems in Thailand. In *Asia's Environmental Crisis,* edited by M.C. Howard. Boulder, CO: Westview Press, 257–274.

Krause, L.B. 1988. Hong Kong and Singapore: Twins or Kissing Cousins? *Economic Development and Culture Change* 36(3): Supplement S45–S66.

Lall, S. 1992. Technological Capabilities and Industrialization. *World Development* 20(2): 165–182.

Lampton, D.A. 1992. A Plum for a Peach: Bargaining, Interest, and Bureaucratic Politics in China. In *Bureaucracy, Politics, and Decision-making in Post-Mao China,* edited by K.G. Leiberthal and D. Lampton. Berkeley: University of California Press, 33–57.

Laughlin, J., and L. Corson. 1995. A Market-Based Approach to Fostering P2. *Pollution Prevention Review* 11–16.

Lee, K.E. 1978. It's Precious and Can No Longer Be Taken for Granted. *New Straits Times* 1: 51.

Lee, K.H. 1986. Malaysia: The Structure and Causes of Manufacturing Protection. In *The Political Economy of Manufacturing Protection: Experiences of ASEAN and Australia*, edited by C. Findlay and R. Gaurnot. Sydney: Allen and Unwin, 99–134.

Lee, K.Y. 2000. *From Third World to First, The Singapore Story: 1965–2000.* New York: HarperCollins Publishers.

Lee, S.Y. 1978. Business Elites in Singapore. In *Studies in ASEAN Sociology,* edited by P.S.J. Chen and H.D. Evers. Singapore: Chopmen Enterprises, 38–60.

Lee, Y.F., and A.Y. So, eds. 1999. *Asia's Environmental Movements: Comparative Perspectives.* Armonk, NY: M.E. Sharpe.

Leonard, J. 1988. *Pollution and the Struggle for the Global Product*. Princeton, NJ: Princeton University Press.

Liddle, R.W. 1991. The Relative Autonomy of the Third World Politician: Soeharto and Indonesian Development in Comparative Perspective. *International Studies Quarterly* 35: 403–427.

Lieberthal, K.G. No date. China's Governing System and Its Impact on Environmental Policy Implementation. Working paper. Washington, DC: Woodrow Wilson Center.

———. 1992. Introduction: The Fragmented Authoritarianism Model and Its Limitations. In *Bureaucracy, Politics, and Decision-making in Post-Mao China*, edited by K.G. Leiberthal and D. Lampton. Berkeley: University of California Press, 1–29.

Lim, C. 1977. Choking Rivers: Discharge from Factory a Threat to 500 Fishermen. *New Straits Times*, September 24, 39.

Lim, I. 1995. Environment in the News. In *Environment and the City*, edited by O.G. Ling. Singapore: Institute of Policy Studies and Times Academic Press, 304–312.

Lim, L. 1983. Singapore's Success: The Myth of the Free Market Economy. *Asian Survey* 23(6): 752–764.

Limanon, W. 1999. Thai Industrial Environmental Performance Rating Program: Using Public Disclosure to Reduce Emissions from Industrial Estates. Master's thesis, George Washington University Law School, Washington, DC.

Lin, S.C. 1997. Experience of a Small Island State in Siting Industrial Plants and Pollutive Facilities. *Journal of Marine and Coastal Law* 12(2): 245–263.

Ling, O.-G. 1994. Centralized Approach in Environmental Management: The Case of Singapore. *Sustainable Development* 2, part 1: 17–22 [United Kingdom].

———. 1995. *Environment and the City—Sharing Singapore's Experience and Future Challenges*, 3d edition. Singapore: Institute of Policy Studies and Times Academic Press.

Lohani, B. 1998. *Environmental Challenges in Asia in the 21st Century*. Manila: Asian Development Bank.

Lohmann, L. 1991. Peasants, Plantations And Pulp: The Politics of Eucalyptus in Thailand. *Bulletin of Concerned Asian Scholars* 23(4): 3–17.

Lovei, M., and C. Weiss. 1997. *Environmental Management and Institutions in the OECD Countries: Lessons from Experience*. Environment Department. Washington, DC: World Bank.

Low, L. 1993. The Economic Development Board. *Challenge and Response: Thirty Years of the Economic Development Board*. Singapore: Times Academic Press, 61–120.

Lung-chu, C., and H.D. Lasswell. 1967. *Formosa, China, and the United Nations*. New York: Saint Martin's Press.

MacIntyre, A. 1993. The Politics of Finance in Indonesia: Command, Confusion, and Competition. In *The Politics of Finance in Developing Countries*, edited by S. Haggard, C.H. Lee, and S. Maxfield. Ithaca, NY: Cornell University Press, 123–164.

———. 1994. Power, Prosperity, and Patrimonialism: Business and Government in Indonesia. In *Business and Government in Industrializing Asia*, edited by A. MacIntyre. Ithaca: Cornell University Press, 244–267.

———. 2001. Institutions and Investors: The Politics of the Economic Crisis in Southeast Asia. *International Organization* 55(1): 81–122.

Mackie, J., and A. MacIntyre. 1994. Politics. In *Indonesia's New Order: The Dynamics of Socio-Economic Transformation,* edited by H. Hill. Sydney: Allen and Unwin, 1–53.

Mardon, R. 1990. The State and the Effective Control of Foreign Capital. *World Politics* 43: 111–138.

Mathews, J.T. 1997. Power Shift. *Foreign Affairs* 76(1): 50–66.

McCormick, B.L., S. Su, and X. Xiao. 1992. The 1989 Democracy Movement: A Review of the Prospects for Civil Society in China. *Pacific Affairs* 65(2): 182–202.

McDowell, M.A. 1989. Development and the Environment in ASEAN. *Pacific Affairs* 62(3): 307–325.

McElroy, M.B., C.P. Nielson, and P. Lydon, eds. 1998. *Energizing China: Reconciling Environmental Protection and Economic Growth.* Cambridge, MA: Harvard University Press.

Means, G.P. 1998. Soft Authoritarianism in Malaysia and Singapore. In *Democracy in East Asia,* edited by L. Diamond and M.F. Plattner. Baltimore: Johns Hopkins University Press, 96–112.

Mehmet, O. 1986. *Development in Malaysia.* London: Croom Helm.

Melby, J.F. 1968. Origins of the Cold War in China. *Pacific Affairs* 41(1): 19–33.

Morell, D., and S. Chai-anan. 1981. *Political Conflict in Thailand.* Cambridge, MA: Ockgeschlager, Gunn and Hain.

MOSTE (Ministry of Science, Technology, and the Environment, Thailand). 1997. *Report of the Industrial Estate Environmental Performance Enhancement Committee.* Bangkok: MOSTE.

———. 1998. *PCD Industrial Estate Monitoring Report.* Bangkok: MOSTE.

Murray, D. 1996. The 1995 National Elections in Thailand: A Step Backward for Democracy. *Asian Survey* 36(4): 361–375.

Muscat, R. 1994. *The Fifth Tiger.* Tokyo: United Nations University Press.

Nelson, K. 1994. Funding and Implementing Projects that Reduce Waste. In *Industrial Ecology and Global Change,* edited by R. Socolow, C. Andrews, E. Berkhout, and V. Thomas. Cambridge, U.K.: Cambridge University Press, 371–382.

Nelson, R.R., ed. 1993. *National Innovation Systems: A Comparative Analysis.* New York: Oxford University Press.

NEPA (National Environmental Protection Agency, China). 1996. *Annual Environmental Yearbook.* Beijing: NEPA.

O'Connor, D. 1994. *Managing the Environment with Rapid Industrialization.* Paris: Organisation for Economic Co-operation and Development.

Ockey, J. 1993. Chaopho: Capital Accumulation and Social Welfare in Thailand. *Crossroads* 8(1): 48–77.

———. 1994. Political Parties, Factions, and Corruption in Thailand. *Modern Asian Studies* 28(2): 251–277.

Office of Science and Technology Advisors. 1995. *A Cleaner Home and a Better Image Abroad: Taiwan's Environmental Efforts.* Taipei: Taiwan Environmental Protection Administration.

Okabe, T. 1998. China's Prospects for Change. In *Democracy in East Asia,* edited by L. Diamond and M.F. Plattner. Baltimore: Johns Hopkins University Press, 171–183.

Pack, H. 1994. Productivity or Politics: The Determinants of the Indonesian Tariff Structure. *Journal of Development Economics* 44: 441–451.

Pagiola, S. 1999. Deforestation and Land Use Changes Induced by the East Asia Crisis. Draft paper. Environment Department. Washington, DC: World Bank.

Panayotou, T. 1998. The Effectiveness and Efficiency of Environmental Policy in China. In *Energizing China: Reconciling Environmental Protection and Economic Growth,* edited by M.B. McElroy, C.P. Nielson, and P. Lydon. Cambridge, MA: Harvard University Press, 431–472.

Panayotou, T., and C. Zinnes. 1994. Free Lunch Economics for Industrial Ecologists. In *Industrial Ecology and Global Change*, edited by R. Socolow, C. Andrews, F. Berkhout, and V. Thomas. Cambridge, U.K.: Cambridge University Press, 383–397.

Payne, R.A. 1995. Freedom and the Environment. *Journal of Democracy* 6(3): 41–55.

PCD (Pollution Control Department, Singapore). 1980. *1980 Pollution Control Report.* Singapore: Environmental Policy and Management Division, Ministry of the Environment.

———. 1994. *1994 Pollution Control Report.* Singapore: Environmental Policy and Management Division, Ministry of the Environment.

Pletcher, J. 1991. Regulation with Growth: The Political Economy of Palm Oil in Malaysia. *World Development* 19(6): 623–636.

Porter, M., and C. van der Linde. 1995. Toward a New Conception of the Environment–Competitiveness Relationship. *Journal of Economic Perspectives* 9(4): 97–118.

Punyaratabandhu, S. 1998. Thailand in 1997. *Asian Survey* 38(2): 161–168.

Quigley, K. 1996. Environmental Organizations and Democratic Consolidation in Thailand. *Crossroads* 9(2): 1–29.

Ramayah, J. 1979. A Bit of Bother in Semenyih Village. *New Straits Times*, October 31, 14.

Ramsay, A. 1985. Thai Domestic Politics and Foreign Policy. Paper prepared for Third United States–ASEAN Conference, January 7–11, Chiangmai, Thailand.

Rashid, A. 1979. Dream Catch Turns into Nightmare for Fishermen. *New Straits Times*, November 1, 4.

Reardon-Anderson, J. 1997. *Pollution, Politics, and Foreign Investment in Taiwan: The Lukang Rebellion.* Armonk, NY: M.E. Sharpe.

Repetto, R. 1988. *The Forest for the Trees: Government Policies and Misuse of Forest Resources.* Washington, DC: World Resources Institute.

Rhee, Y., B. Ross-Larson, and G. Pursell. 1984. *Korea's Competitive Edge: Managing Entry into World Markets.* Baltimore: Johns Hopkins University Press.

Riggs, F. 1966. *Thailand: The Modernization of a Bureaucratic Polity.* Honolulu: East–West Center Press.

Roberts, M., and J. Tybout, eds. 1996. *Industrial Evolution in Developing Countries.* New York: Oxford University Press.

Robison, R. 1986. *Indonesia and the Rise of Capital.* Sydney: Allen and Unwin.

Rock, M.T. 1994. Transitional Democracies and the Shift to Export-Led Industrialisation: Lessons from Thailand. *Studies in Comparative International Development* 29(1): 18–37.

———. 1995. Thai Industrial Policy: How Irrelevant Was It to Export Success? *Journal of International Development* 7(5): 745–757.

———. 1996a. Industry and the Environment in Ten Asian Countries: Synthesis Report of the US–AEP Country Assessments. Paper prepared for United States–Asia Environmental Partnership (AEP), October 9, Washington, DC.

———. 1996b. Toward More Sustainable Development: The Environment and Industrial Policy in Taiwan. *Development Policy Review* 14(3): 255–272.
———. 1999. Re-Examining the Role of Industrial Policy in Indonesia: Can the Neo-Liberals Be Wrong? *World Development* 27(4): 691–704.
———. 2000a. Globalization and Sustainable Industrial Development in the Second Tier Southeast Asian Newly Industrializing Countries. Paper prepared for *Global Change Assessment Report*, supported by Asia Pacific Network for Global Change Research. Chiang-Mai, Thailand: Impact Center for Southeast Asia.
———. 2000b. Thailand's Old Bureaucratic Policy and Its New Semi-Democracy. In *Rents and Rent-Seeking and Economic Development: Theory and the Asian Evidence,* edited by M. Khan and K.S. Jomo. Cambridge, U.K.: Cambridge University Press, 183–206.
———. 2000c. Using "Green Taxes" to Increase Revenues and Improve Environmental Management in Local Government Following Decentralization. Paper prepared for U.S. Agency for International Development–Indonesia under the Natural Resources Management Program, Washington, DC.
———. 2000d. Selective Intervention in Agriculture on the Road to Industrialization in Indonesia, Malaysia, and Thailand. Under review by *Journal of International Development.*
Rock, M.T., O.-G. Ling, and V. Kimm. 2000. Public Policies to Promote Cleaner Shared Industrial Growth in East Asia. In *Asia's Clean Revolution*, edited by D.P. Angel and M.T. Rock. Sheffield, England: Greenleaf Publishing, 88–103.
Rodan, G. 1989. *The Political Economy of Singapore's Industrialization.* Kuala Lumpur: Macmillan Press.
Roht-Arriaza, N. 1995. Shifting the Point of Regulation: The International Organization for Standardization and Global Lawmaking on Trade and the Environment. *Ecology Law Quarterly* 22(3): 479–539.
Roome, N., ed. 1998. *Sustainable Strategies for Industry.* Washington, DC: Island Press.
Ross, L. 1988. *Environmental Policy in China.* Bloomington: Indiana University Press.
Ruzecki, I. 1997. *Strategy for the Use of Market-Based Instruments in Indonesia's Environmental Management.* Office of Environment and Social Development. Manila: Asian Development Bank.
Salleh, I.M., and M. Meyanathan. 1993. *The Lessons of East Asia: Malaysia, Growth, Equity, and Structural Transformation.* Washington, DC: World Bank.
Sani, S. 1993. *Environment and Development in Malaysia.* Kuala Lumpur: Center for Environmental Studies at Institute for Strategic and International Studies.
Scott, J.C. 1985. *Weapons of the Weak: Everyday Forms of Peasant Resistance.* New Haven, CT: Yale University Press.
Seagrave, S. 1985. *The Soong Dynasty.* New York: Harper & Row.
Shamsul, A.B. 1983. The Politics of Poverty Eradication: The Implementation of Development Projects in a Malaysian District. *Pacific Affairs* 53(3): 455–476.
Silcock, T.H. 1985. *A History of Economics Teaching and Graduates in Singapore, 1934–1960.* Department of Economics and Statistics. Singapore: University of Singapore.
Silva, E. 1997. Democracy, Market Economics and Environmental Policy in Chile. *Journal of Inter-American Studies and World Affairs* 38(4): 1–33.
Sinkule, B., and L. Ortolano. 1995. *Implementing Environmental Policy in China.* Westport, CT: Praeger.

Smil, V. 1993. *China's Environmental Crisis: An Inquiry into the Limits of National Development*. Armonk, NY: M.E. Sharpe.

Snodgrass, D.R. 1980. *Inequality and Economic Development in Malaysia*. New York: Oxford University Press.

So, A.Y., and Y.F. Lee. 1999. Environmental Movements in Thailand. In *Asia's Environmental Movements: Comparative Perspectives*, edited by Y.F. Lee and A.Y. So. Armonk, NY: M.E. Sharpe, 120–142.

Somsak, S.L. 1997. Toxic Fallout Haunts Mae Moh Villagers. *Bangkok Post*, June 30. /bkkpost/1997/june1997/bp970630/30006_news06.html (accessed November 2001).

———. 1998. Prosecute Power Plant Call. *Bangkok Post*, June 16. /bkkpost/1998/june1998/bp19980616/160698_news08.html (accessed November 2001).

———. 1999. Water Near Mae Moh Plant Found to Be Contaminated. *Bangkok Post*, March 7. /bkkpost/1999/march1999/bp19993007/070399_news.html (accessed November 2001).

Spofford, W., X. Ma, J. Zou, and K. Smith. 1996. Assessment of the Regulatory Framework for Industrial Pollution Control in Chongqing. Working paper. Washington, DC: Resources for the Future.

Steering Committee. 1989. *Taiwan 2000: Balancing Economic Growth and Environmental Protection*. Taipei: Republic of China.

Sterner, T. Forthcoming. *Policy Instruments for Environmental and Natural Resource Management*. Washington, DC: Resources for the Future.

Sunderlin, W.D. 1999. Between Danger and Opportunity: Indonesia's Forests in an Era of Economic Crisis and Political Change. www.cgiar/research/projects/project1/crisis1.html (accessed November 2001).

Sunee, M., and J. Canino. 1998. Environmental Law in Thailand. In *Comparative Environmental Law and Regulation*, edited by N. Robinson. Dobbs Ferry, NY: Oceana Publications, THL-3–THL-21.

Suphavit, P. 1996. Thailand Country Experiences with the Use of Economic Instruments. Paper presented at Asia–Pacific Economic Cooperation Expert Meeting on Innovative Approaches Towards Environmentally Sustainable Development, Quezon City, the Philippines, June 6–7.

Suthy, P. 1982. *The Nature of Thai Business and Implications for U.S. Investors*. Bangkok: Faculty of Economics, Chulalongkorn University.

Tamin, M., and S. Meyanathan. 1988. Rice Market Intervention System in Malaysia: Scope, Effects and the Need for Reform. In *Evaluating Rice Market Interventions: Some Asian Examples*. Manila: Asian Development Bank, 91–150.

Tan, H.A., and G. Batra. 1995. *Enterprise Training in Developing Countries: Incidence, Productivity Effects, and Policy Implications*. Private Sector Development Department. Washington, DC: World Bank.

Tang, S., and C.P. Tang. 1997. Democratization and Environmental Politics in Taiwan. *Asian Survey* (37)3: 281–294.

———. 1999. Democratization and the Environment: Entrepreneurial Politics and Interest Representation in Taiwan. *China Quarterly* 158 (June): 350–366.

Tay, J. 1993. Environmental Protection in Singapore. Working paper. Singapore: Nanyang Technical University.

TEPA (Taiwan Environmental Protection Administration). 1993. *1993: State of the Environment Taiwan, R.O.C.*. Taipei: TEPA.

———. 2000. *National Air Quality Improvement Results.* Taipei: TEPA, 1.

Times Academic Press. 1993. *Challenge and Response: Thirty Years of the Economic Development Board.* Singapore: Times Academic Press.

Timmer, C.P. 1975. The Political Economy of Rice in Asia: Indonesia. *Food Research Institute Studies* 14(3): 197–231.

———. 1989. Agricultural Prices and Stabilization Policy. Development discussion paper 290 AFP. Cambridge, MA: Harvard Institute for International Development.

———. 1993. Rural Bias in the East and Southeast Asian Rice Economies. *Journal of Development Studies* 30(2): 149–176.

———. 1996. Does Bulog Stabilize Rice Prices? Should It Try? *Bulletin of Indonesian Economic Studies* 32(2): 45–74.

TISI (Thailand Industrial Standards Institute). 2000. *ISO 14000 Certification: Tally and Details*. Bangkok: TISI.

Tsong-Juh, C. 1994. A Review of Present Status of Laws and Regulations on Environmental Protection in the Republic of China. In *Taiwan 2000: Balancing Economic Growth with Environmental Protection.* Taipei: Republic of China, 438–441.

UNEP (United Nations Environment Program). 1997. *Engaging Stakeholders: The 1997 Benchmark Survey on Company Environmental Reporting.* Paris: UNEP.

Vermeer, E.B. 1998. Industrial Pollution in China and Remedial Policies. *China Quarterly* 156 (April): 952–985.

Vincent, J.R. 1993. *Reducing Effluent While Raising Affluence: Water Pollution Abatement in Malaysia.* Cambridge, MA: Harvard Institute for International Development.

Vincent, J.R., M. Rozali, and associates. 1997. *Environment and Development in a Resource Rich Economy.* Cambridge, MA: Harvard University Press.

Vincent, J.R., M.A. Rozali, and A.R. Khalid. 2000. Water Pollution Abatement in Malaysia. In *Asia's Clean Revolution,* edited by D. Angel and M. Rock. Sheffield, U.K.: Greenleaf Publishing, 173–193.

Vogel, D. 1986. *National Styles of Regulation.* Ithaca, NY: Cornell University Press.

Wade, R. 1990. *Governing the Market.* Princeton, NJ: Princeton University Press.

———. 1999. Greening the World Bank: The Struggle over the Environment, 1970–1995. In *The World Bank: Its First Half Century, Vol. 2: Perspectives,* edited by D. Kapur, J.P. Lewis, and R. Webb. Washington, DC: Brookings Institution, 611–734.

Walder, A.G. 1992. Local Bargaining Relationships and Urban Industrial Finance. In *Bureaucracy, Politics, and Decision-Making in Post-Mao China,* edited by K.G. Leiberthal and D. Lampton. Berkeley: University of California Press, 308–333.

Wang, H., and B. Lui. 1999. Policy-Making for Environmental Protection in China. In *Emerging China: Reconciling Environmental Protection and Economic Growth,* edited by M.B. McElroy, C.P. Nielson, and P. Lydon. Cambridge, MA: Harvard University Press, 371–404.

Wang, H., and D. Wheeler. 1996. *Pricing Industrial Pollution in China: An Econometric Analysis of the Levy System.* Policy Research Department and Environmental Division. Washington, DC: World Bank.

———. 1999. Endogenous Enforcement and the Effectiveness of China's Pollution Levy System. Paper presented at workshop on Market-Based Instruments for Environmental Protection. Harvard University, Cambridge, MA, July.

Warnick, I., R. Herman, S. Govind, and J. Ausubel. 1996. Materialization and De-Materialization: Measures and Trends. *Daedalus* 125(3): 171–198.

Wasant, T. 1999. Villagers Won't Be Relocated. *Bangkok Post*, February 22. /bkkpost/1999/february1999/bp19990.../220299_news03.html (accessed November 2001).

Wheeler, D., and P. Martin. 1992. Prices, Policies and the International Diffusion of Clean Technology: The Case of Wood Pulp Production. In *International Trade and the Environment,* edited by P. Low. Washington, DC: World Bank, 197–224.

World Bank. 1989. *Indonesia: Forest, Land, and Water, Issues in Sustainable Development.* Report 7822-IND. Washington, DC: World Bank.

———. 1990. *World Tables: 1989–90.* Baltimore: Johns Hopkins University Press.

———. 1992a. *China: Environmental Strategy Paper. Vols. I and II.* Report 9669-CHA. Washington, DC: World Bank.

———. 1992b. *World Development Report 1992.* New York: Oxford University Press.

———. 1993. *Malaysia: Managing the Costs of Urban Pollution.* Report 11764-MA. Washington, DC: World Bank.

———. 1994a. *Indonesia: Environment and Development.* Washington, DC: World Bank.

———. 1994b. *Thailand: Mitigating Pollution and Congestion in a High Growth Economy.* Report 11770-TH. Washington, DC: World Bank.

———. 1994c. *World Development Report 1994.* New York: Oxford University Press.

———. 1996a. *Managing Capital Flows in East Asia.* Washington, DC: World Bank.

———. 1996b. *World Development Report 1996.* New York: Oxford University Press.

———. 1997a. *Can the Environment Wait? Priorities for East Asia.* Washington, DC: World Bank.

———. 1997b. *Clear Water, Blue Skies.* Washington, DC: World Bank.

———. 1998. *World Development Indicators.* Washington, DC: World Bank.

———. 2000a. *Greening Industry.* New York: Oxford University Press.

———. 2000b. *Thailand Environmental Monitor 2000.* Washington, DC: World Bank.

World Commission on Environment and Development. 1987. *Our Common Future.* Oxford: Oxford University Press

Yun-han, C. 1998. Taiwan's Unique Challenges. In *Democracy in East Asia,* edited by L. Diamond and M.F. Plattner. Baltimore: Johns Hopkins University Press, 133–146.

Index

About the Author

Michael T. Rock is chair of the Department of Economics and Management at Hood College. His previous positions include senior economist at the Winrock International Institute for Agricultural Development and dean of faculty at Bennington College. Rock's published research focuses on trade and the environment, deforestation in poor countries, the environmental behavior of manufacturing plants in first- and second-tier newly industrializing economies (NIEs) of East Asia, water policy, and the role of the state in Southeast Asia's second-tier NIEs. He has taught and lived in Thailand and Vietnam. He is coauthor (with James Weaver and Kenneth Kusterer) of *Achieving Broad-Based Sustainable Development* and coeditor (with David Angel) of *Asia's Clean Revolution: Industry, Growth, and the Environment.*